Lecture Notes in Biomathematics

Managing Editor: S. Levin

Brain Theory Subseries

63

Carme Torras i Genís

Temporal-Pattern Learning in Neural Models

Springer-Verlag
Berlin Heidelberg New York Tokyo

Author

Carme Torras i Genís
Institut de Cibernètica (UPC-CSIC)
Diagonal 647, 2ª planta, 08028 Barcelona, Spain

Mathematics Subject Classification (1980): 92; 34C, 58F, 70K

ISBN 3-540-16046-9 Springer-Verlag Berlin Heidelberg New York Tokyo
ISBN 0-387-16046-9 Springer-Verlag New York Heidelberg Berlin Tokyo

Printing and binding: Beltz Offsetdruck, Hemsbach/Bergstr.
2146/3140-543210

PREFACE

While the ability of animals to learn rhythms is an unquestionable fact, the underlying neurophysiological mechanisms are still no more than conjectures. This monograph explores the requirements of such mechanisms, reviews those previously proposed and postulates a new one based on a direct electric coding of stimulation frequencies. Experimental support for the option taken is provided both at the single neuron and neural network levels.

More specifically, the material presented divides naturally into four parts: a description of the experimental and theoretical framework where this work becomes meaningful (Chapter 2), a detailed specification of the pacemàker neuron model proposed together with its validation through simulation (Chapter 3), an analytic study of the behavior of this model when submitted to rhythmic stimulation (Chapter 4) and a description of the neural network model proposed for learning, together with an analysis of the simulation results obtained when varying several factors related to the connectivity, the intraneuronal parameters, the initial state and the stimulation conditions (Chapter 5).

This work was initiated at the Computer and Information Science Department of the University of Massachusetts, Amherst, and completed at the Institut de Cibernètica of the Universitat Politècnica de Catalunya, Barcelona. Computers at the latter place have adopted Catalan as their mother tongue and thus some computer-made figures in this monograph, specially those in Chapter 5, appear labeled in that tongue.

"*El pensament s'afirma i s'aferma en les objeccions: doneu-me un bon contradictor i seré capaç d'inventar les més excelses teories*", J. Fuster dixit. Without claiming the excellence of the theory, I am however greatly indebted to M.A. Arbib, L. Basañez, W. Buño and J. Fuentes for their support and enlightening discussions.

Carme Torras

Barcelona, April 1985

TABLE OF CONTENTS

CHAPTER 1
INTRODUCTION

ACHILLES: "I wonder if it would be possible for me to learn all about my neural structure -so much so that I would be able to predict the path of my neural flash before it even covered its path! Surely, this would be total, exquisite self-knowledge".

TORTOISE: "Oh, Achilles, you have innocently thrown yourself into the wildest of paradox".

D.R. Hofstadter, 1981

Living beings with nervous systems can use experience to modify their behavior in response to certain stimuli. These behavioral changes are mediated by, and can be ultimately reduced to, consistent variations in the firing pattern of some neurons. The experimental discovery that, through repeated stimulation, long-lasting changes can be induced in the firing pattern of certain isolated neurons suggests possible approaches to the analysis of the neural substratum of learning and memory. It is within this framework that the present work has been developed.

1.1 - MOTIVATION AND OBJECTIVES

The neurons' learning of temporal patterns of stimulation, subject of many experimental studies, has scarcely been dealt with from the perspective of neural modelling. This omission, though not especially significant itself, leads to the much more crucial lack of temporal information in the forms of representation postulated as a neural substratum of learning and memory.

The majority of neural learning models have focussed on the spatial aspects both of stimulation and of their own activity, thus shaping the approach and the methods used in this research area, to the point of defining its characteristics, which in turn become its limitations. Three of these characteristics/limitations are:

(1) The use of logical neuron models.

(2) The consideration of the nervous system as a passive stimulus-transducer.

(3) The circumscription of learning to changes in synaptic efficacy.

The first point is both cause and consequence of the fact that meaning is attributed only to the instant-by-instant states, disregarding evolution. Experimental data have revealed more and more shortcomings of the analogy which compares the neuron with a logical device. The complexity of the processing and storage capacities found in neurons has recently caused the microprocessor to be considered as a new candidate for this comparison (Barto and Sutton, 1981b)[£].

In relation to the second point, it has been shown that the nervous system of mammals is constantly active and that a great part of this activity is spontaneous; the stimulation coming from the environment exerts a modulating effect on it, rather than determining it uniquely (Kandel, 1976; Bunge, 1981).

The third point presupposes that learning modifies not the autonomous functioning of neurons, but the influence that some have over others or that the stimuli exert on them, thus only implying structural changes in the network. The most representative exponent of this tendency --the Hebbian rule-- has not obtained the experimental support expected, giving rise to the assertion that its evidence comes from psychology (classical conditioning), rather than from neurophysiology (Levine, 1984).

The research line undertaken in this work intends, as expressed in the title, to address the neural modelling of the learning of temporal patterns, proposing alternatives to overcome the three limitations we have just mentioned. This general aim is concretized in the objective of setting up and analyzing the behavior of a neuron model and of a neural network model based on the former, both able to explain and reproduce, through simulation, a set of electrophysiological phenomena found upon submitting certain isolated neurons and some animals in learning situations to repeated rhythmic stimulation. Thus, we seek

----

(£): These same authors point out the tradition for the most complex invention of each era becoming a metaphor for the functioning of the nervous system. Thus the computer, which was initially proposed as a metaphor for the brain (Arbib, 1972), has recently been replaced by the more modern network of microprocessors.

microscopic mechanisms that might give rise to certain macroscopic phenomena linked to the learning of rhythms.

Our overall objective can be divided into serial sub-objectives, that make up the work-plan we will follow:

(1) Design a biophysical neuron model provided with adaptive capabilities which permit the learning of rhythms.

(2) Validate this model.

(3) Study the model's behavior analytically.

(4) Specify a network model that incorporates the above-mentioned neuron model and whose connectivity is consistent with the experimental data.

(5) Explore, through simulation, the conditions under which the network reproduces the electrophysiological phenomena under consideration.

1.2 - REFERENCE FRAMEWORK

Given the interdisciplinary nature of the subject, we will briefly outline the dominant trends in each of the research areas the present work is related to, following the line of conceptual development that has determined the successive incorporation of each area.

We will start with the **theory of adaptive neural networks**, from which the motivation for this work arises and whose state-of-the-art has been outlined in the previous section. From this discipline, we have taken the method for specifying and categorizing the learning strategies: supervised or unsupervised, open- or closed-loop, and guided by an objective or by a reinforcement signal. We have also taken into account the collected experience regarding the advantages and limitations of the rules previously proposed.

To approach the neural learning of temporal patterns of stimulation, it is essential to consider the available **neurophysiological data obtained from different nervous system formations of vertebrates subjected to rhythmic stimulation in a conditioning paradigm.** Relevant to our work here is the hypothesis that, based on these data, has emerged in the experimental domain: That rhythmic stimuli are coded at the neural level in terms of their own frequency and that, if this is true, other types of stimuli could be stored using a similar code, through spatio-temporal patterns of neural activity (Thatcher and John, 1977). Thus, the mentioned code would serve as a substratum of learning and memory.

To overcome the first limitation expressed in the preceding section, and taking into consideration that the data to model are electrophysiological in nature, we have adopted for the neuron model a level of description (electrical variables) and type of formalism (differential equations) used in the **biophysical modelling of intracellular activity.** This discipline has remained much closer to the experimental domain than the previously outlined theory of adaptive neural networks, having as its primary objective to relate certain functions of the nervous system to its structure on the basis of morphological and physiological evidence.

Let us briefly digress to assert that in the spectrum of domains to which mathematical modelling has been applied, devised by Karplus (1983), according to how much is known about the modelled system, the neurophysiological domain has a position very close to the extreme of the unknown. Each position of the spectrum has a purpose and a factor of validity associated with it. In the case we are dealing with, the purpose is to acquire the knowledge we lack, and the low validity factor suggests that we extract mostly qualitative conclusions (Vibert and Segundo, 1979). It is clear that the aforementioned is no more than an overall tendency, since for some experimental preparations there are very accurate quantitative models, the validity of the unineuronal models being higher, in general, than that of the multineuronal ones.

Continuing on the expository line previously undertaken, we find that, to overcome the second limitation, and in view of the indirect evidence of the existence of spontaneously firing neurons in the nervous systems of vertebrates, we have opted for an oscillator as our neuron model. The nonlinear oscillators traditionally proposed as models of neurons are of two kinds: relaxation and limit-cycle. Specifically, the model proposed in this monograph is of the first kind and some classic results of the **Theory of Oscillators** (Pavlidis, 1973; Winfree, 1980) have been employed to clarify some aspects of its dynamics, when subjected to rhythmic stimulation.

The analytic study of the mentioned dynamics has permitted us to approach a question which, until now, had been left open: the characterization of the patterns of entrainment between oscillators of a specific kind and periodic stimuli, as well as the positions of the entrainment regions within the space of all possible conditions of stimulation (Segundo and Kohn, 1981). To answer this question, we have used the **topological properties of the mappings from the circle onto itself** and, briefly, some tools from **Measure Theory** and from the **Elementary Theory of Numbers.**

In relation to the validation of the pacemaker neuron model, taking into account the impossibility of isolating one neuron from the nervous system of a vertebrate and the similarity of their functioning to the neurons of invertebrates, we have used the **data obtained by submitting some pacemaker neurons of invertebrates to rhythmic stimulation.** Concerning the most relevant aspect of our research --plasticity in the firing pattern-- there are hints in the changes in the intracellular activity brought about by repeated rhythmic stimulation, which allow us to postulate tentative learning rules with enough biological content to suggest experiments that might confirm or refute the suppositions on which they are based.

Finally, contrasting the simulation results with the experimental ones, on the neuron level as well as on the network level, has required the use of a set of **electrophysiological signal processing techniques** (Glaser and Ruchkin, 1976; Fuentes, 1979), especially those that permit the detection of synchronizations and the characterization of firing frequencies.

1.3 - STRUCTURE OF THE TEXT

The reference framework outlined in the preceding section is enlarged upon in the following chapter, where its constituent experimental and theoretical elements are detailed separately. Specifically, the first part is devoted to the description of the set of electrophysiological phenomena that we attempt to model, detailing the data that, on the unineuronal and multineuronal level, will be used afterwards for the validation of the neuron and network models, respectively. The second section reviews the neuron models of autorhythmicity and of learning proposed up to now.

The third chapter contains the specification and the validation of the neuron model proposed. After the listing of the experimental and theoretical requirements that must be satisfied and an informal preliminary description of the model, there is the neurophysiological validation of the assumptions on which it is based and its formal specification. Later, we undertake the a posteriori validation of the model, contrasting the simulation results with the experimental data at the unineuronal level presented in the preceding chapter.

In Chapter 4, an analytic study is made of the patterns of entrainment between a simplified version of the aforementioned model and a rhythmic stimulus. The behavior of the oscillator subjected to the stimulus is characterized by a mapping of the circle onto itself, whose periodic solutions allow determination of the conditions under which entrainment takes place and derivation of the detailed input/output patterns that emerge. Finally, we analyze the effect of randomness and learning upon the mentioned patterns.

Chapter 5 is devoted to the description and simulation of the neural network model. In the first section, the connectivity is defined, experimental evidence is presented that supports it and also justifies the incorporation into the network of the pacemaker neuron model proposed, and the neural network model is formally specified. In the second section, the implementation, as well as the types of measures and graphics used to characterize the behavior of the network, are detailed. Next, there is a report and comments on the simulation results, establishing in the first place a reference experiment, from which the effect that the variation of several factors has upon the behavior of the network is afterwards analyzed.

The sixth and last chapter contains a synthesis of the most outstanding results and contributions of this work, as well as a brief indication of the most promising lines of future research.

Finally, we include three appendices. The first lays out some basic neurophysiological notions, on which the specific data to be modelled, collected in Chapter 2, are based. The second appendix contains a listing of the properties of the mappings from the circle onto itself, used in Chapter 4 to deduce the entrainment conditions and the input/output patterns that emerge. As a third appendix, a glossary of the Catalan labels appearing in computer-made figures is included.

CHAPTER 2
EXPERIMENTAL DATA AND PREVIOUS MODELS

- "Tengo una idea", decía un escritor demasiado reminiscente.
- "¿De quién?", le atajó un amigo.

S. Ramón y Cajal, 1921

Two types of elements --experimental and theoretical-- constitute the reference framework where the objectives set forth and the results obtained from this work become meaningful.

In relation to the experimental elements, based on the generic knowledge of neuron physiology outlined in Appendix A, there is a set of specific data about neurons and neural networks which show rhythmic behavior and have a certain plasticity. This set of data, described in the first section, is what the neuron and network models proposed attempt to explain and reproduce; it therefore constitutes the experimental framework in which the mentioned models must be validated (Zeigler, 1976).

With respect to the theoretical elements, in the second section we review the neuron models previously proposed to account for the phenomena of autorhythmicity and learning. These models provide a series of results assumed in setting up the neuron and network models proposed, together with an indication of the limitations of certain strategies and, above all, a topography of this research field, comprising a compilation of different open questions and an evaluation of their relative importance.

2.1 - ELECTROPHYSIOLOGICAL RECORDINGS AND PHENOMENA TO BE MODELLED

"What we need is a smart electrode".

G.W.Harding i A.L. Towe, 1978

This section describes the set of electrophysiological phenomena that we attempt to model and that will be taken as reference when undertaking the a posteriori validation of the proposed models.

Other data that support certain options that have been taken, concerning intraneuronal mechanisms and the structure of the network, will be laid out in Sections 3.2.2 and 5.1.2, where we will discuss the a priori validation of the neuron and network models, respectively.

2.1.1 - Data at the unineuronal level

There are two kinds of recordings at this level: intracellular and extracellular. The first, which requires the insertion of a microelectrode through the neuron membrane, shows the evolution of the difference of potential between the interior of the cell and the extracellular milieu (Figure 2.1(a)). The second, obtained by putting two electrodes on the axon surface, shows the instants in which the neuron fires (Figure 2.1(b)).

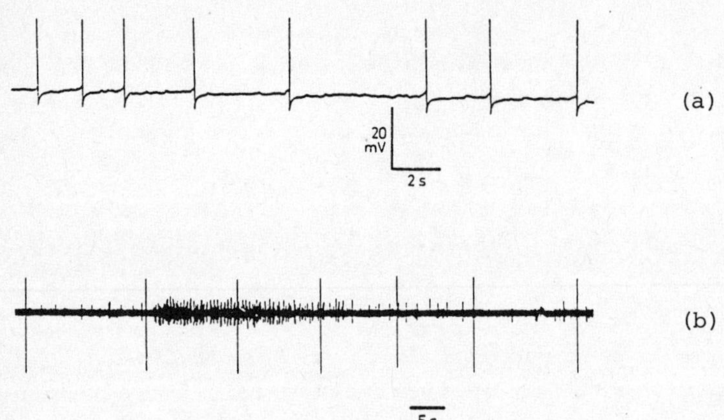

*** Figure 2.1 - Types of unineuronal recordings: (a) intracellular, (b) extracellular.

The phenomena that we try to model are those reflected in the extracellular recordings obtained under certain conditions, or in graphics derived from a statistical treatment of these data. We will consider, in particular, the histogram of interspike intervals, the phase response curve and the curve that represents the firing period as a function of the stimulation period.

The data that we will present correspond to different pacemaker neurons belonging to the nervous systems of several invertebrates(£), among

----

(£): squid, sea snail, oyster, watersnake, sea lobster, crayfish, medicinal leech, limax, and land lobster.

which we will point out those located in the abdominal ganglion of the sea snail, Aplysia. To these neurons, which in the last two decades have been the subject of intensive research, we have adjusted the parameters of the model proposed in Chapter 3.

2.1.1.1 - Variability of the spontaneous interspike interval

Rhythmic neurons are not ideal oscillators. Instead, their spontaneous interspike intervals vary from cycle to cycle (Figure 2.2). In the literature on this subject, this variability is usually characterized through the coefficient of variation of the histogram of spontaneous interspike intervals, because of the wide range of average frequencies appearing even in neurons of the same type belonging to different individuals of the same species.

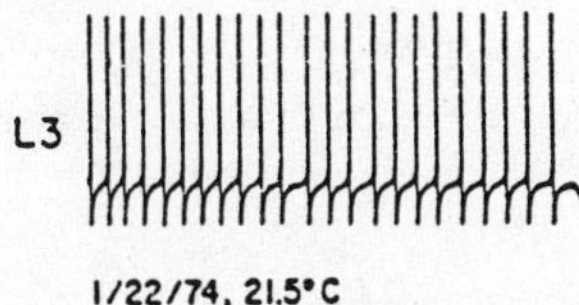

*** Figure 2.2 - Variability of the spontaneous interspike interval in Aplysia neurons (from Pinsker and Kandel, 1977).

A representative case is the tonic stretch receptor in the last thoracic segment of crayfish (Procabarus clarkii), which, for each stretching level maintained constant, shows an average firing frequency that varies from individual to individual between 2 and 15 Hz, with a coefficient of variation between 0.02 and 0.1 (Buño and Fuentes, 1984, 1985). Similar data had been obtained previously by Firth (1966), Bustamante (1980), Kohn (1980) and Kohn et al. (1981).

Frequency bands and coefficients of variation of the same order have been reported for neurons of the abdominal ganglion of Aplysia (Junge and Moore, 1966) and the watersnake (Lymnaea stagnatis) (Holden and Ramadan, 1980).

In relation to the shape of the histograms of spontaneous interspike intervals (Figure 2.3), several authors concur in stating that they are unimodal and slightly positively skewed (Junge and Moore, 1966; Tuckwell, 1978; Holden and Ramadan, 1980, 1981b; Kohn et al., 1981).

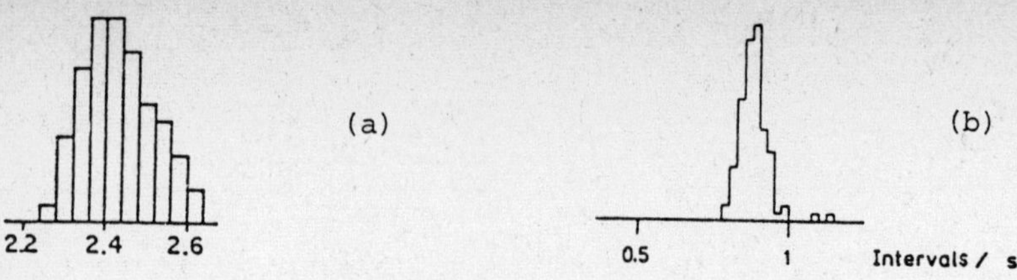

*** Figure 2.3 - Histograms of spontaneous interspike intervals: (a) for neurons of Aplysia (from Junge and Moore, 1966); (b) for a neuron of the watersnake (Lymnaea stagnatis) (from Holden and Ramadan, 1981b).

2.1.1.2 - Response to occasional perturbations. The Phase Response Curve.

We understand by the **phase** $\emptyset$ of an oscillator, the fraction of the period elapsed since the last occurrence of the event that indicates the beginning of the period. In the case of pacemaker neurons, this event is the discharge and as period T_0 we consider the average spontaneous interspike interval. We can thus write:

$$\emptyset = \frac{T}{T_0}$$

where T is the time elapsed since the previous discharge.

Occasional perturbations usually have only an ephemeral effect on the period of an oscillator, but a permanent effect on the phase.

The **Phase Response Curve (PRC)**, frequently used in theoretical and experimental studies on oscillators, describes the aforementioned permanent effect that an occasional input impulse has on the response of the oscillator. More precisely, the PRC is a graphical representation of the delay (or advancement) of phase provoked by the stimulus, as a function of the phase of the oscillator just before the arrival of the stimulus.

Figure 2.4 shows the PRCs obtained experimentally for some neurons of the abdominal ganglion of Aplysia and the stretch receptor of crayfish.

The two PRCs for excitatory stimuli can be approximated by two straight regression lines in the form of a "V", while the ones for inhibitory stimuli can be roughly approximated by a single straight line of slope between 0 and 1.

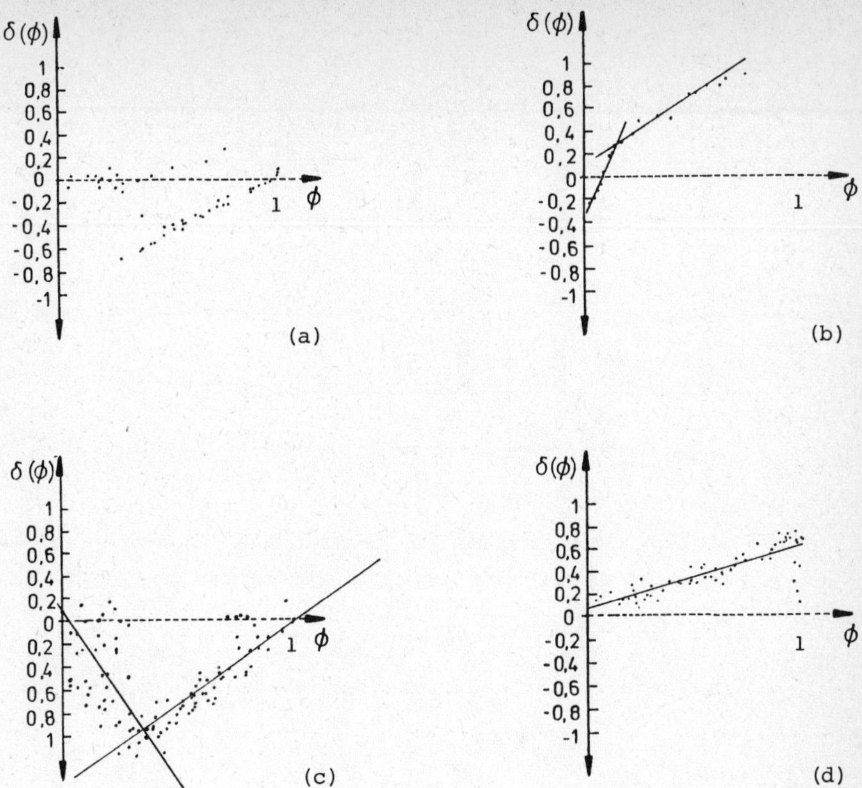

*** Figure 2.4 - Phase response curves: (a) and (c) for excitatory stimuli; (b) and (d) for inhibitory stimuli; (a) and (b) for neurons of the abdominal ganglion of Aplysia (from Pinsker and Ayers, 1983; and Pinsker, 1977a); (c) and (d) for the stretch receptor of crayfish (from Buño and Fuentes, 1985; and Kohn et al., 1981). $\delta(\emptyset)$ is the phase-delay provoked by an input impulse, as a function of the phase $\emptyset$ of the oscillator just before the arrival of that impulse. Thus, negative values in the ordinate axis mean phase-advancement.

Except for the PRCs obtained for cardiac neurons, which have a substantially different shape from that in Figure 2.4 (Jalife and Antzelevitch, 1979; Guevara et al., 1981; Peterson and Calabrese, 1982), the rest of the PRCs reported in the literature for other pacemaker neurons are qualitatively similar to the ones represented in the mentioned figure (Ayers and Selverston, 1977a, 1979; Beltz and Gelperin, 1980; Pinsker and Ayers, 1983).

It should be mentioned that the PRC $\delta(\emptyset)$ vs. $\emptyset$ that we use in the present work is one of the three types of curve that appear in the literature under the name of PRC. The other two represent the co-phase (fraction of the period that remains to complete the cycle after the incidence of the stimulus) and the new phase (phase in which the oscillator is left after the incidence of the stimulus) as a funtion of $\emptyset$. Figure 2.5 is an attempt to clarify these terms.

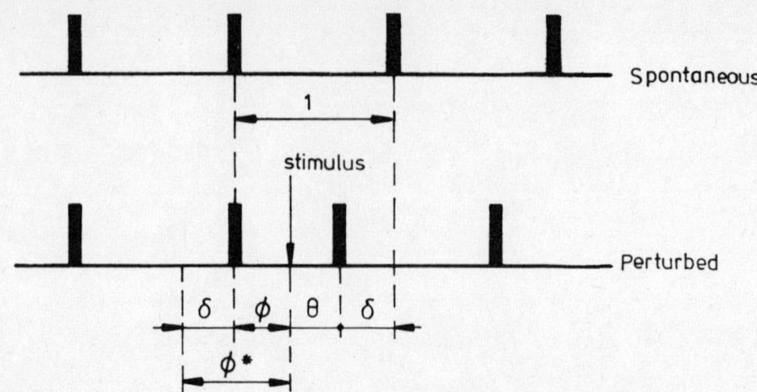

*** Figure 2.5 - Meaning of the terms phase ($\emptyset$), delay or advancement of phase (δ), co-phase (θ) and new phase ($\emptyset^*$).

Buño and Fuentes (1985) point out that these three kinds of curve are related by simple geometric transformations. In effect, given that:

$$\emptyset + \delta + \theta = 1$$
$$\emptyset^* + \theta = 1$$

we need only apply to $\delta(\emptyset)$ a counterclockwise rotation of 45º in relation to the point (1,0), to obtain $\theta(\emptyset)$. $\emptyset^*(\emptyset)$, on the other hand, results from reflecting $\delta(\emptyset)$ in relation to the axis of abscissas and then applying a counterclockwise rotation of 45º with respect to the origin (Figure 2.6).

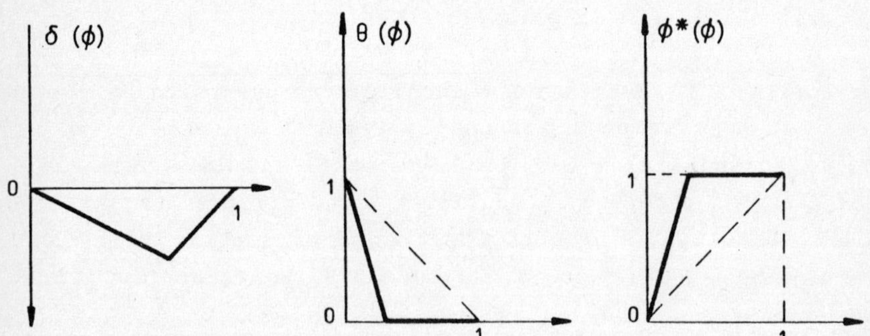

*** Figure 2.6 - Geometric relationships between: $\delta(\emptyset)$, $\theta(\emptyset)$ and $\emptyset^*(\emptyset)$.

The PRC is a mapping from the circle into the circle; thus, it is representable on the torus. If it is continuous on this surface, we can assign to it a rotation number: number of circles gone through by the dependent variable for each circle gone through by the independent variable (Appendix B). Winfree (1980) points out that, of the infinite rotation numbers that could theoretically appear, only two --0 and 1-- have actually done so in experimental preparations, for which reason that author classifies oscillators into two groups: Type 0 and Type 1[£]. Until now, there have been no theoretical explanations given for this experimental finding.

The PRC of certain oscillators varies according to the amplitude of the stimulation (Figure 2.7) (Pinsker, 1977b; Ayers and Selverston, 1979; Buño and Fuentes, 1985). There are neurons that under low-amplitude stimuli behave as oscillators of Type 1 and, once a certain threshold of stimulation amplitude is exceeded, they behave as Type 0 oscillators (Jalife and Antzelevitch, 1979; Guttman et al., 1980; Winfree, 1980; Peterson and Calabrese, 1982; Buño and Fuentes, 1984, 1985). In those cases in which the PRC is continuous in relation to the stimulus amplitude (m), besides being so in relation to the phase, the application of the unit disc into the circle presents necessarily a singularity (point in which the advancement of phase is ambiguous and at whose surroundings it can take any value), according to the topological properties of the continuous mappings with the domain and image mentioned above.

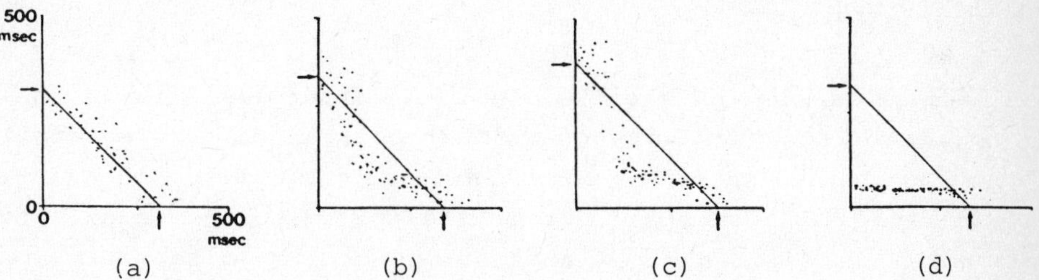

*** Figure 2.7 - Variation of the PRC with increase --from (a) to (d)-- of the stimulus amplitude (from Buño and Fuentes, 1985).

These questions will be dealt with in Section 3.3.2, referring to the neuron model proposed. The PRC, apart from being taken into account at

----

(£): We must note that Winfree uses PRCs of the type $\emptyset^*(\emptyset)$ and that the rotation numbers 0 and 1 for this curve correspond to the numbers -1 and 0 for the curve $\delta(\emptyset)$.

the time of validating the model a posteriori, will be used in Section 3.3.3 to calculate the bands of stimulation frequencies that yield entrainment and in Section 4.1 to derive the phase transition equation.

In the previous paragraph, the two latter uses of the PRC presuppose the validity of the following conditions:

(a) The phase-shift produced by the arrival of an input impulse is independent of the number of input impulses applied within an inter-spike interval.

(b) The arrival of an input impulse modifies the duration of the present interval, but does not affect subsequent intervals.

In sum, both uses require that the PRC always and completely characterizes the effect that an input impulse has on the phase of the neuron.

Though in a preliminary approximation the reality seems to adjust to these ideal conditions (Kohn, 1980; Kohn et al., 1981), a more detailed study reveals certain divergences. Research efforts are being devoted to characterizing the circumstances in which these divergences arise, as well as the form they adopt in each case (Buño and Fuentes, 1984, 1985). Until these results are known, the researchers who have studied the entrainment phenomenon analytically, as we do in this thesis, have assumed conditions (a) and (b) to be true (Segundo and Kohn, 1981).

2.1.1.3 - Entrainment

The response of an oscillator to a periodic input usually contains two preponderant frequency components that correspond to its intrinsic rhythm and to the input. For some nonlinear oscillators submitted to stimulation frequencies similar to the intrinsic one or one of its harmonics, the first of the mentioned components disappears and we obtain a response at the stimulation frequency or one of its divisors. The described phenomenon is called one-to-one or subharmonic entrainment, according to whether the frequency is the intrinsic one or a divisor.

In general, we call entrainment the relationship between an oscillator and a periodic input where the sequence of phases of the oscillator in the moments of arrival of the impulses is periodic. **(s:r) entrainment** refers to the fact that s cycles of the stimulus correspond to r cycles of the response (Figure 2.8). Synchronization is a particular case, since it requires the coincidence of both null phases.

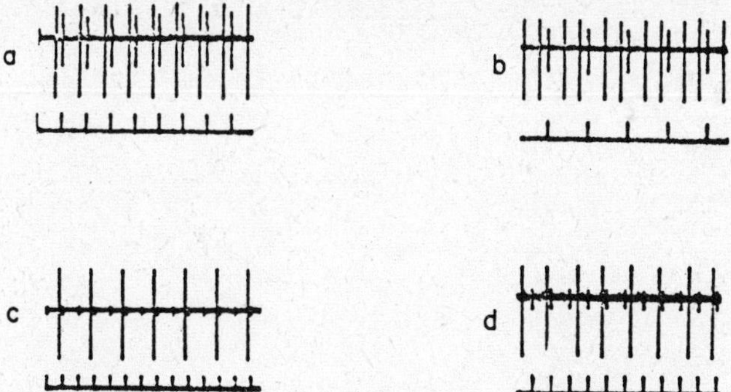

*** Figure 2.8 - Several entrainment ratios. (a) 1:1, (b) 1:2, (c) 2:1, (d) 3:2 (from Kohn et al., 1981).

Entrainment is a frequently-observed phenomenon in experiments with pacemaker neurons subjected to periodic stimulation, either excitatory or inhibitory (Perkel et al., 1964; Segundo and Perkel, 1969; Stein, 1974; Kristan and Calabrese, 1976; Pinsker, 1977b; Ayers and Selverston, 1977a, 1979; Eberly et al., 1979; Hartline and Gassie, 1979; Kristan, 1980; Holden and Ramadan, 1981a; Kohn et al., 1981; Pinsker and Ayers, 1983). In the representation of the firing period as a function of the stimulation period (Figure 2.9), the bands of input frequencies that yield the different entrainment ratios appear as rectilinear segments aligned with the origin. The overall trend of the distribution of points is increasing for the excitatory stimuli and decreasing for the inhibitory ones.

We next enumerate the most salient features, as reported by different experimenters, that characterize entrainment in invertebrate pacemaker neurons:

(a) The bands of stimulation frequencies that cause entrainment widen as the amplitude of the stimulation increases (Pinsker, 1977b; Pinsker and Ayers, 1983).

(b) For excitatory inputs, the entrainment is almost or completely synchronized (the co-phase is constant and almost null). For inhibitory inputs, the entrainment is never synchronized (Ayers and Selverston, 1979).

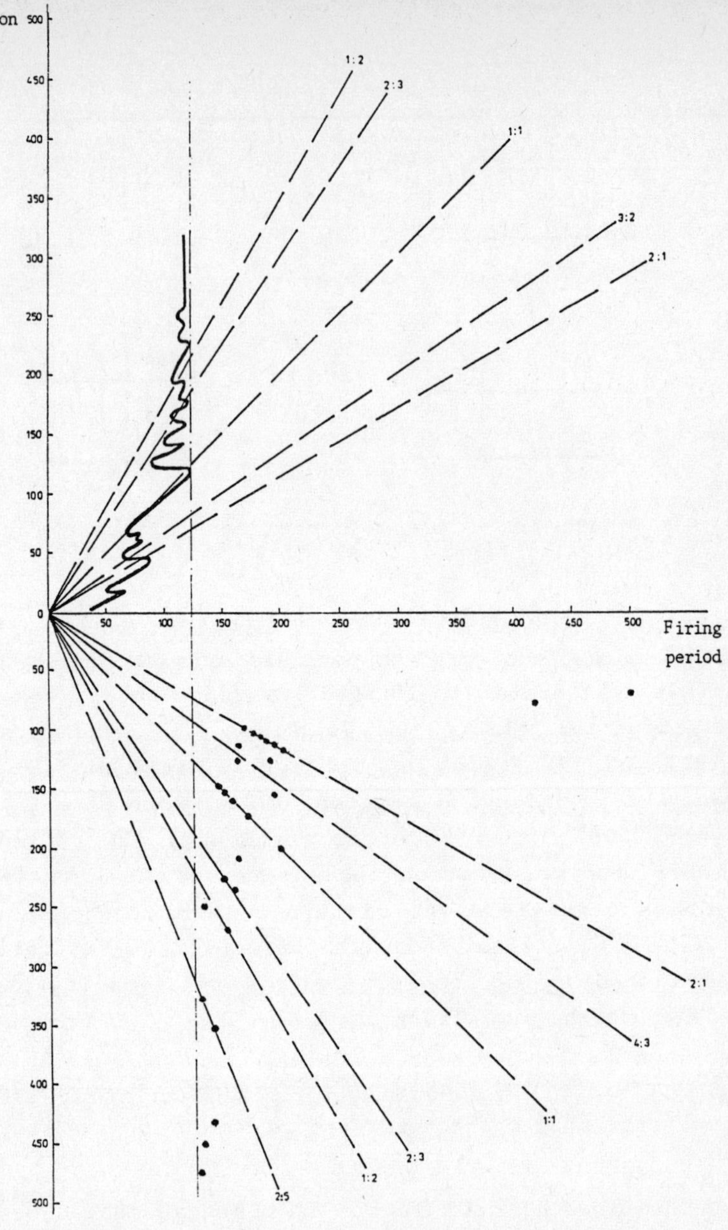

*** Figure 2.9 - Representation of the firing period as a function of the stimulation period, for pacemaker neurons of Aplysia. The rays indicate the location in the graph of some simple entrainment ratios. The curve in the upper quadrant is the result of interpolating the data obtained by applying excitatory stimulation at several frequencies (from Moore et al., 1963). The dots in the lower quadrant are a sample of the results obtained by applying inhibitory stimulation (from Kohn et al., 1981).

(c) If the ratio between the stimulation and firing periods gives place to stable entrainment, this is attained after a few cycles, regardless of what the initial phase was. Whether unstable entrainment is obtained depends, on the contrary, on the initial phase (Pinsker, 1977b).

(d) For high frequencies of inhibitory stimulation, the firing frequency decreases rapidly toward zero (Kohn et al., 1981).

(e) While some authors maintain that entrainment always arises (Kohn et al., 1981), others appeal to the variability of the neuron's endogenous oscillatory activity to explain the appearance of aperiodic responses (Holden and Ramadan, 1981a).

(f) Most researchers have predominantly observed responses with simple entrainment ratios (1:1, 2:1, 1:2, 3:1,...), which is usually justified also by the intrinsic variability of the neuron (Holden and Ramadan, 1981a). The above explanation becomes more plausible when there is evidence that the mentioned responses are the least sensitive to noise incorporated into the stimulation (Kohn et al., 1981).

(g) Entrainment is, in general, more regular for excitatory than for inhibitory stimulation. In the excitatory case, it is maximum for frequencies slightly higher than the spontaneous one, and in the inhibitory case, it is maximum for frequencies slightly lower than the spontaneous one (Pinsker, 1977b; Ayers and Selverston, 1979).

2.1.1.4 - Plasticity in the firing pattern

Neuronal plasticity refers to the consistent changes of certain properties of the neuron as a function of the stimulation and its own previous activity. Despite the close relationship between them, we usually distinguish between morphologic and physiologic plasticity. The first is predominant in the nervous systems of individuals that are still in a developmental stage. The second, on the contrary, is found also in adults and is the one that is considered the substratum of behavioral learning.

Several electrophysiological techniques have permitted showing that the firing patterns of a neuron can be modified for minutes, hours, or even days, through specific strategies of stimulation. These strategies can be classified according to two criteria: (1) Whether they are associative, and (2) whether they use natural or artificial stimulation (direct electrical stimulation).

The non-associative strategies were the first to be applied at the cellular level, probably because they require only one channel of stimulation. The most widely reported phenomena made evident by this procedure have been: post-tetanic potentiation, synaptic depression, sensitization and habituation. All of them amount to an increase or decrease in the amplitude of the postsynaptic potential induced by a stimulus, as a result of its repeated application. Since the model we propose is not oriented to the exploration of phenomena which, like the previous ones, require intracellular recordings for their study, but will consider only the specific plasticity shown by pacemaker neurons subjected to periodic stimulation, observable from extracellular recordings, we simply note that a compilation of the experimental results relative to the above-listed phenomena can be found in Wine and Krasne (1978).

In relation to the kind of specific plasticity that we try to reproduce with the proposed model, von Baumgarten (1970) points out that a pacemaker neuron of the abdominal ganglion of Aplysia, under excitatory stimulation of slightly higher or lower frequency than the spontaneous one, for a period between a few minutes and an hour, continues firing at the imposed frequency even after the stimulation has stopped. This behavior may last only a few cycles (Figure 2.10) or as long as 20 minutes.

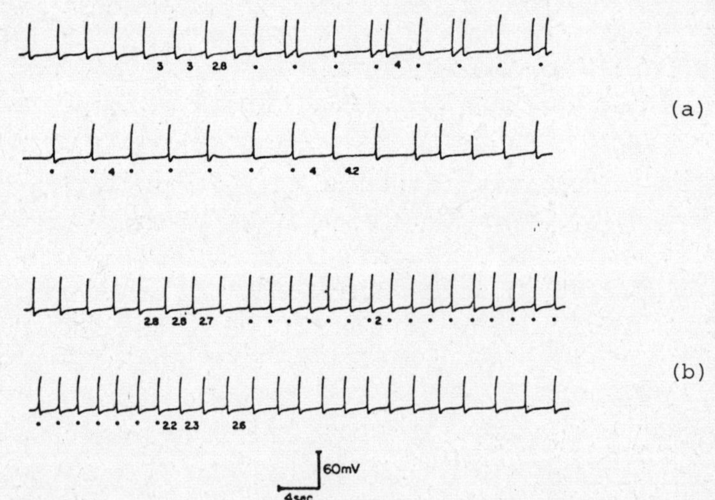

*** Figure 2.10 - Plasticity in the firing pattern: (a) lengthening, and (b) shortening of the interspike interval, resulting from the application of rhythmic excitatory stimuli to neurons in the abdominal ganglion of Aplysia (from von Baumgarten, 1970). Dots indicate the instants of presentation of input impulses.

Because of its relevance, we will include here a datum that affects the a priori validation of the model and, therefore, will be commented on in Section 3.2.2. In the experiment described above, the spontaneous membrane potential of the pacemaker neuron adjusts progressively to the requirements of the stimulus, increasing its rate of growth from one period to another, if the frequency of stimulation is higher than the spontaneous one, and diminishing it in the opposite case.

Similar results have been obtained with bursting neurons of the same ganglion (von Baumgarten, 1970; Parnas et al., 1974). In this case, plasticity affects not only the interburst interval, but also the number of spikes within a burst and the interspike interval; the effects last for much longer time periods of up to several hours.

We will include in what follows some of the results derived from the use of associative strategies which, though not belonging to the strict experimental framework that we try to model, are directly related to it, since they support the possibility of varying the spontaneous firing rate of a pacemaker neuron by means of external stimulation.

Associative strategies are, for example, classical and instrumental conditioning. The first requires the pairing of two stimuli and the second, the pairing of spontaneous activity and reinforcement. The use of such strategies with artificial stimulation at the cellular level has been especially espoused by Kandel (1968, 1976), who pointed out the advantage of using the periodic discharges of pacemaker neurons as spontaneous activity in the instrumental conditioning paradigm. It has been observed that the pairing of two stimuli can lead to a greater variation in the firing rhythm than the sum of those induced by each one separately. This phenomenon is known as "heterosynaptic facilitation" or "heterosynaptic inhibition", depending on the direction of the variation (Kristan, 1971). An increment in the firing frequency can also be attained through negative reinforcement (Tosney and Hoyle, 1977; Woolacott and Hoyle, 1977). Other experiments with bursting neurons show it is possible to vary the number of spikes in each burst, the interspike interval and the interburst interval, stimulating the neuron in a way contingent upon its phase (Pinsker and Kandel, 1977).

One higher-level aspect that we will not deal with at this time, but that because of its importance should be mentioned, is the extrinsic control of intrinsic plasticity. Apart from including plastic neurons, nervous systems have mechanisms to control the development, expression and maintenance of the potential changes in those neurons. This allows them to prevent changes that would not be adaptative. Krasne (1978) offers a documented summary of these mechanisms.

2.1.2 - Data at the multineuronal level

"There are no facts, only interpretations"

F. Nietzsche, 1901

"El paisaje es un estado del alma"

E. Sábato, 1953

Two types of recordings are usually employed to study the electrical activity in wide groups of neurons: the electroencephalogram (EEG) and evoked potentials. The former simultaneously provides a continuous monitoring of the endogenous activity in different regions of the cortex, without reference to any external stimulus, while the latter show the average response of a region to successive presentations of the same stimulus.

By means of these techniques, a great deal of evidence has been obtained indicating that learning at the behavioral[£] level is linked to long-lasting changes in the firing pattern of differentiated groups of neurons. Some of these changes are also inducible using artificial strategies of stimulation. Several surveys of the different approaches taken by the research in this area, as well as of the results obtained, are available (Thompson and Patterson, 1972; Leiman and Christian, 1973; Buser, 1976; Rosenzweig and Bennet, 1976; Thatcher and John, 1977; Teyler, 1978; Mc Fadden, 1979; Woody, 1982).

Our interest is centered in a very specific portion of these results: the ones obtained using the tracer technique. This technique consists of presenting an intermittent stimulus (visual, auditory, electrical) to an animal in a learning situation and following the trace of its characteristic frequency in the recordings that correspond to different regions in the brain of the animal. This technique, introduced by

----

(£): Considering that all the experiments described in this section refer to behavioral learning, the expression "conditioned response" in this context will denote an animal behavior.

Livanov and Poliakov (1945), has been widely used by E.R. John and his colleagues (John, 1967, 1972; Ramos et al., 1976a & b; Thatcher and John, 1977; John and Schwartz, 1978).

Below, we briefly describe the phenomena that we attempt to model, as reported by this research team:

(a) **Progressive entrainment with a rhythmic stimulus:** When an animal is subjected to a conditioning process where the conditioned stimulus (CS) is rhythmic, initially the EEG only appears entrained to the stimulus in peripheral regions tied to the pertinent sensory modality (in the visual case, the optic nerve and the lateral geniculate nucleus), for entrainment to spread to other more central unspecific regions (visual cortex, reticular formation, hippocampus) and become more marked, as conditioning progresses and more meaning is attached to the CS (Figure 2.11).

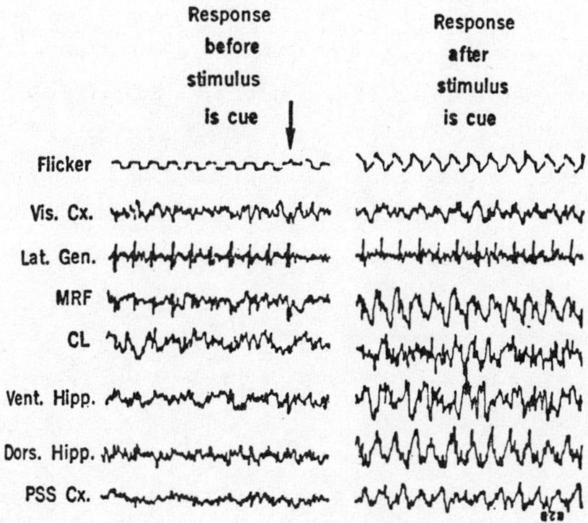

*** Figure 2.11 - Response of different parts of the nervous system of a cat to a rhythmic stimulus, before and after it acquires the meaning of a conditioned stimulus (from John, 1972).

(b) **Prediction of the time of presentation of the stimulus:** When entrainment with the stimulus in central regions has been attained and the stimulus stops, a characteristic positive wave occurs at the moment when the input impulse should have arrived.

(c) **Assimilation of the rhythm:** While the animal remains in the conditioning situation, it is observed that in the successive intertrial periods the EEG becomes more and more dominated by the frequency of the stimulus or one of its harmonics.

(d) **Generalization:** The presentation of the stimulus at a slightly different frequency from that of the CS triggers the conditioned response and determines the predominance in the EEG of the two characteristic frequencies: the presented one (in the peripheral regions) and the previously learned one (in the central regions).

(e) **Discrimination:** When two responses are conditioned to two different stimulation frequencies, the range of generalization of each of them diminishes. If the stimulus is presented at an intermediate frequency, both this frequency and the two previously learned are present in the EEG and, of the latter two, the one predominating is that associated with the response emitted.

(f) **Relationship between the evoked rhythms and unineuronal activity:** About 30% of the neurons of the lateral geniculate nucleus and the visual cortex vary their firing rhythms as a result of conditioning. The detailed firing pattern of some of them becomes a perfect replica of the stimulus, even after turning it off (a phenomenon called "assimilation of the rhythm at the cellular level" by some authors). In the discrimination experiments, an analysis of the individual firing patterns reveals the appearance of only the two learned frequencies, as well as a third one if a testing stimulus has been presented.

The described facts have been observed by several researchers working with rats, rabbits, cats, dogs, monkeys and humans, using a large variety of experimental designs.

What is especially thought-provoking in all these phenomena is less the ability shown by certain neural structures to reproduce the electrical activity previously induced by a stimulus, in the absence of it, than the fact that the stimulus' frequency itself provides a code for the external events, on which memory and learning of rhythms rely. Many other experiments, using different strategies of stimulation, support the above statement:

(a) **Cortical conditioning:** The repeated pairing of a constant tone with a flickering light determines the appearance of electrical activity at the frequency of the light stimulus, when only the tone is presented.

(b) **Instrumental conditioning:** The quantity and quality of entrainment between the EEG and a periodic stimulus depends on the patterns of reinforcement used.

(c) **Spontaneous appearance of a response previously conditioned to a rhythmic stimulus:** This response is preceded by electrical activity at the frequency of the CS in different brain regions.

(d) **Transfer of learning:** Transfer of a response initially conditioned to an auditory intermittent stimulus, to a visual conditioned stimulus presented at the same frequency as the auditory one, is accomplished much faster than the transfer of the same response to a constant tone. Thus, equality of frequency preponderates over equality of sensory modality.

(e) **Mistaken response in a discrimination experiment:** the coincidence between the dominant frequencies in the EEGs of the peripheral and central regions, observed when the response is correct, disappears when the response is erroneous (Figure 2.12).

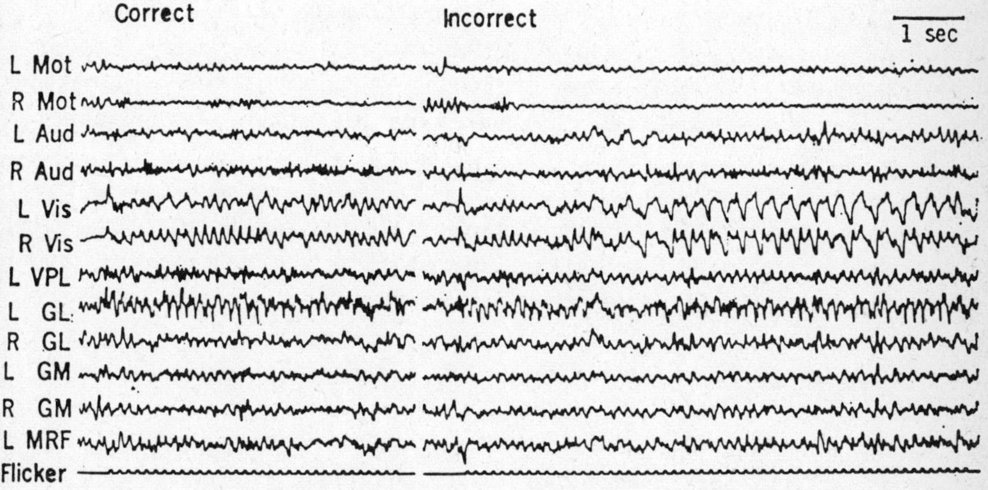

*** Figure 2.12 - Comparison between the electrophysiological activities elicited, in several regions of the nervous system of a cat, by a single stimulus interpreted in two different ways: one correct and the other incorrect (from John, 1972).

(f) **Differential conditioning:** If the meaning of a rhythmic CS becomes progressively more explicit and precise, the distribution, duration and precision of the frequency assimilated in the EEG increases. The CS can be made progressively more explicit, for example, through the following stages: simple conditioning, differentiation between a frequency associated with a response and another associated with a non-response, discrimination between two frequencies associated with two different responses, etc.

(g) **Analysis of conflict situations:** Two CSs, of different frequencies and different sensory modalities, associated with incompatible conditioned responses, are simultaneously presented. The frequency predominant in the regions not tied to any sensory modality is the one associated with the emitted response.

(h) **Conditioning of animals with a segmented corpus callosum:** Since these animals have separate brain hemispheres, it is possible to use one hemisphere as a control, so that the effects of the repeated presentation of a stimulus become clearly differentiated from those resulting from learning.

All the experiments described have something in common: The use of intermittent stimuli, the trace of which can be seen in the electrical activity of different brain regions. There is evidence that learning, in different experimental situations that involve stimuli of this kind, is based on the electric coding of the mentioned stimuli in terms of their frequency. Can this electric coding in terms of temporal patterns of activity be generalized to other types of stimuli? The answer to this question requires the use of evoked potentials, since the consistent changes in the electrical activity are not, in general, directly detectable. Through the mentioned potentials, it has been observed that the average electrical activity subsequent to the presentation of any CS varies consistently as conditioning progresses.

The peculiarity of the electric coding in the experiments that make use of the tracer technique is transparency: frequencies are coded as frequencies. This peculiarity disappears when using other kinds of stimuli, but the most fundamental fact, that learning and memory are at least partly based on an electric endogenously-reproducible coding of the external events, remains.

2.2 - PREVIOUS MODELS

"Physical models are as different from the world as a geographical map is from the surface of the earth".

L. Brillouin

"... y los Colegios de Cartógrafos levantaron un Mapa del Imperio que tenía el Tamaño del Imperio y coincidía puntualmente con él".

J.L. Borges, 1954

The neuron and network models that we propose in the present work establish a point of connection between two areas that have traditionally remained far apart: the biophysical modelling of intracellular activity and the theory of adaptive neural networks. The models proposed in the first area are of the continuous-time type and use variables that correspond to electrical and biochemical magnitudes, directly measureable experimentally, to reproduce as faithfully as possible the biological reality. The strategies of neuronal adaptation, on the other hand, have usually been formulated on logical neurons of the linear-addition-with-threshold type, and their study, carried out always at the network level, has been oriented toward determining the restrictions imposed by certain structures and the general principles of neurophysiological functioning that can serve as a substratum to the different kinds of behavioral learning.

The maintenance of the separation between those two research areas, except for some few cases (Lara et al., 1976, 1980), is explained by the conjunction of three tendencies:

(a) The predominance of hypotheses that, like the Hebbian one, attribute plasticity only to synaptic efficacy, and whose implementation, therefore, does not require a detailed intracellular modelling.

(b) The focussing of biophysical modelling on aspects of neural functioning more primary than plasticity, which still remain unknown.

(c) The complexity of most biophysical models proposed, which makes it very expensive in computing time to simulate a network of minimal dimensions and prohibits analytic treatment.

We will maintain the double panorama when outlining the framework in which this work is situated, describing in Section 2.2.1 the pacemaker neuron models proposed, all biophysical in nature and, in Section 2.2.2, the neuronal models of learning.

2.2.1 - Models of pacemaker neurons

The different nonlinear oscillators which have been proposed as models of pacemaker neurons can be joined into two general classes: relaxation oscillators and limit-cycle oscillators. The first ones presuppose a variable that increases (or decreases) monotonically and that, when it surpasses a threshold, is automatically reset to its initial value. The second ones, on the contrary, require a minimum of two state-variables and are characterized by the existence of a closed convergent trajectory in their phase-plane. An exhaustive and detailed description of these and other types of nonlinear oscillators is found in Minorsky (1962) and Andronov et al. (1966).

Many of the models we are considering come from the modification of some parameter conditions in the models previously established for non-pacemaker neurons; in the same way as many models of the relaxation oscillator class have, in turn, been obtained by simplification of models of the limit-cycle type. The triplet Hodgkin-Huxley (1952) / BVP (1961) / Perkel et al. (1964) constitutes a paradigmatic instance of this simplifying process. The first of these models has become a true landmark in the evolution of neural modelling, since it reproduces with such exactness the flow of sodium and potassium ions underlying the action potential, and so has been used as reference in the validation of many later models. The great value of this model, which incorporates four state-variables, resides in having established a correspondence between experimental data about the same phenomenon, obtained at two very different levels (evolution of the action potential and changes in the permeability to sodium and potassium). The BVP (Bonhoeffer / van der Pol) model, proposed by Fitzhugh, reduces the state-variables of the previous model to two, and shows in its phase-plane a limit-cycle attained for a certain level of depolarization. Finally, the model proposed by Perkel et al. supposes a further simplification, since, in it, the phase characterizes the system uniquely, so that any perturbation is translated into a phase-shift, as in all relaxation oscillators.

We will not describe the non-pacemaker neuron models here and, among the pacemaker ones, we will highlight in the exposition only the BVP, the Radial Isochron Clock and those proposed by Perkel et al. and by Hartline. Two good compilations of the first kind of models, separated by an interval of a decade, are those provided by Harmon and Lewis

(1966) and McGregor and Lewis (1977). Concerning the second ones, Pavlidis (1973), Winfree (1980) and Pinsker and Ayers (1983) constitute essential references.

2.2.1.1 - Limit-cycle oscillators

The BVP model results of combining Bonhoeffer's model of the excitability of iron wire in nitric acid (Bonhoeffer, 1948) with van der Pol's model of the heartbeat (van der Pol, 1926). Its analytic expression is:

$$\dot{x} = c\left(y + x - \frac{x^3}{3} + z\right)$$

$$\dot{y} = -\frac{(x - a + by)}{c} \tag{2.1}$$

where the variables x and y are approximately linear functions of two different pairs of state-variables of the Hodgkin-Huxley (H-H) model, z is the input, and the parameters $a,b,c \in \mathbb{R}$ are subject to the following restrictions:

$$1 - \frac{2b}{3} < a < 1$$

$$0 < b < 1 \tag{2.2}$$

$$b < c^2$$

These restrictions ensure the existence of a unique singular point (where the lines $\dot{x}=0$ and $\dot{y}=0$ cross) in the phase-plane (Figure 2.13(a)). The system remains in this point until an input impulse moves it along the dotted line of the previous figure. The state then follows the trajectory that begins in the corresponding point of that line, to return eventually to the resting point. The action potential occurs only for long translations along the dotted line.

A sufficiently-intense constant depolarizing current makes the singular point unstable and surrounded by a stable limit-cycle (Figure 2.13(b)). It is then that the model becomes an oscillator.

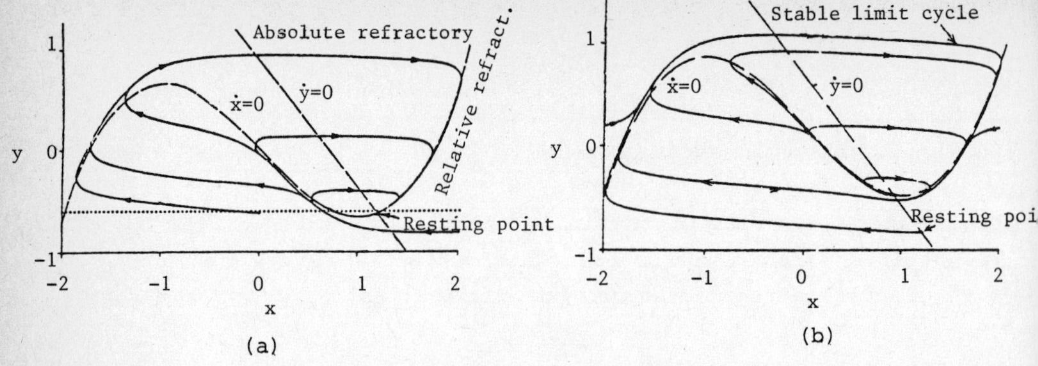

*** Figure 2.13 - Phase-plane representation of the BVP model: (a) in the absence of stimulation (from Harmon and Lewis (1966)); (b) when submitted to a constant depolarizing current. See text for details.

The fact that the BVP model is a representative, with minimum number of state-variables, of the class to which the H-H model belongs has been crucial for understanding the dynamics of this last model. Conversely, the experimental concordance of the H-H model has made it possible to interpret the BVP model in physiological terms, despite its not being based on experimental data and its state-variables not having a direct biological interpretation, which has to be inferred from the functional relationship that they maintain with the ones of the H-H model.

Since the discovery that the H-H model presents a rhythmic behavior for some specific parameter values, many experimenters have introduced small modifications into this model in order to adapt it to different kinds of pacemaker neurons: sensory (Teorell, 1971) and cardiac (McAllister et al., 1975; Bristow and Clark, 1982), mainly.

Nevertheless, when the object of the research has been the neuron's response to periodic stimulation, the above kind of models has been discarded because of the high computational cost the numeric integration of their differential equations requires and the difficulty of obtaining a clear vision of their dynamics under these stimulation conditions. Then simpler models have been used, which we will next describe.

The Radial Isochron Clock (RIC) model is defined, in polar coordinates, by the following differential equations:

$$\dot{\phi} = 2\pi$$

$$\dot{r} = ar(1-r) \quad \text{with} \quad a \in \mathbb{R}^+ \qquad (2.3)$$

Several authors (Guevara and Glass, 1982; Hoppensteadt and Keener, 1982) have studied the behavior of this oscillator submitted to periodic perturbations and maintain that, though biologically unrealistic, the model explains certain aspects of the generation of cardiac arhythmias. Winfree (1980) defined a generalization of this oscillator in which ϕ depends on r. The implications of introducing this modification with regard to the response to rhythmic stimulation have not yet been analyzed.

2.2.1.2 - Relaxation oscillators

A RIC oscillator, with a high-enough value of the parameter "a" (equation 2.3) is practically characterized by the state-variable ϕ and can therefore be assimilated to a clock(£). Neu (1979) showed how to carry out this same assimilation in the most general case of an n-dimensional oscillator with a strongly-attractive limit-cycle.

In the H-H model and its many derivatives, the state-variables vary slowly, except at the time of firing, when the variation becomes extremely quick, without losing continuity. This has favored the development of a great number of models in which the quick transition has been substituted for a discontinuous jump. All these models adhere to the same interpretation of neuronal functioning: The membrane potential increases as a result of the entrance of sodium ions into the cell body, while the threshold for a sudden increment of the permeability to sodium remains constant or decreases gradually from the time of firing; as soon as the values of the two coincide, an action potential is generated and both are discontinuously reset to widely separate initial values, from which they resume their approximation towards equality. Depending on whether the increment in membrane potential is spontaneously driven or requires external stimulation to surpass the threshold, the modelled neuron will or will not be a pacemaker.

(£): A clock is defined by a single state-variable ϕ of constant derivative and is topologically equivalent to a relaxation oscillator.

What are the consequences of the loss of continuity in the response of the model? Basically, the disappearance of the singular variety in the phase-space which, without solution of contiguity, made possible the transition between Type 0 and Type 1 oscillatory behaviors. This kind of singularity in the H-H model was demonstrated analytically by Best (1979), who noted the difficulty of experimentally distinguishing it from a discontinuity.

The general expression of the models that incorporate the mentioned discontinuity, called "integrate-and-fire models", is:

$$\begin{cases} \dfrac{dP}{dt} = F_1(P,x(t)) \wedge \dfrac{dH}{dt} = F_2(H), \text{ while } P<H \\ \\ P=P_o \wedge H=H_o, \text{ when } P=H \end{cases}$$

where P is the membrane potential, H is the threshold, and x(t) is the input.

As we have said, we will limit ourselves to the previous models that show spontaneous oscillatory behavior, but not without first referring briefly to some works done on non-oscillatory models, especially when they become oscillators because of an intense-enough constant stimulus, as the H-H model does.

The simplest instance of non-oscillatory model --the linear integrator-- is that in which $F_1=x(t)$ and $F_2=0$ (Sokolove, 1972; Knight, 1972). Two slightly more sophisticated instances are the leaky-integrator (Rescigno et al., 1970; Knight, 1972; Stein et al., 1972; Keener, 1981) and the integrator with variable time-constant (Fohlmeister et al., 1974, 1977). In both cases, $F_1=-\gamma P+x(t)$, with γ constant in the former and a function of time in the latter. The frequency response of the above models to different intensities of stimulation has been widely studied and characterized in the mentioned papers, as has also the one obtained with discrete versions of those and other related models (Nagumo and Sato, 1971; Sato, 1972; Yoshizawa, 1982; Yoshizawa et al., 1982).

Focussing now on the intrinsically oscillatory models that have been proposed according to the interpretation of the neuronal functioning given before, let us say that they are necessarily of the relaxation oscillator type and differ basically in the functional form attributed to the membrane potential and the threshold evolution with respect to time. The linear, sinusoidal and exponential functions are the most commonly used.

Perkel et al.'s model (1964), proposed in relation to some experimental data about the response of the stretch receptor of crayfish and the abdominal ganglion of Aplysia to periodic stimulation, postulates exponential evolutions in opposite directions for the spontaneous membrane potential and the threshold. The postsynaptic potentials are assumed to be punctate and are instantaneously added to the spontaneous potential, to afterwards drop exponentially to zero. When there is no stimulation, the model behaves as a leaky-integrator with exponential threshold, subjected to a constant input. The exponential evolution of the spontaneous potential is supported by the experimental data of Junge and Moore (1966). These and other researchers (Walloe et al., 1969) have worked on models that suppose slight modifications of Perkel et al.'s model, such as considering the threshold to be constant.

Hartline's model (1976), called the "active pacemaker model", attempts to constitute a generic model, with enough flexibility to adapt to different experimental preparations, either of a single neuron (the stretch receptor of crayfish; Hartline, 1976a & b) or of a network with few neurons (the stomatogastric ganglion of the sea lobster; Hartline, 1979). The works done with this model especially emphasize the need to measure experimentally the parameters, instead of adjusting them to reproduce the output patterns observed. To forecast in the design of the model the later incorporation of measurements specifically related to new experimental situations, three adaptation factors and one conductivity have been included, in addition to the classical variables: membrane potential and threshold. The membrane potential has not only been divided into a spontaneous component and a postsynaptic component, but in the spontaneous one a distinction has also been made between an active and a passive part, both following exponential trajectories with different time-constants, of which the first is so large that its corresponding trajectory nears linearity. The way to accumulate the postsynaptic potentials is a compromise between addition and integration; this gives the model a fast-response capacity to impulsional stimuli and another, slower one, to continuous stimulation. In most simulations carried out, constant values have been assigned to the adaptation and the conductivity factors, since, according to Hartline himself, the model is not necessarily the simplest that can reproduce, with a specific degree of precision, the results obtained in a limited set of experiments. The model betters the previous ones in the reproduction of the detailed trajectory followed by the membrane potential in the stretch receptor of crayfish.

Other relaxation oscillators have been proposed as generic models in relation to several oscillatory phenomena, among them the neural ones. We would like to highlight the models proposed by Glass and Mackey (1979) and by Enright (1980). The former assumes a linear membrane potential and a sinusoidal threshold, and the latter, a constant membrane potential and a linear threshold.

Finally, some authors interested less in the internal functioning of pacemaker neurons that in their response to perturbations, have further simplified the models, characterizing the neurons by their PRC (Segundo, 1979; Segundo and Kohn, 1981). These models are useful when studying the neuron's response to rhythmic stimulation.

2.2.2 - Neuronal models of learning

The theory of adaptive neural networks originated with, and continues to be strongly influenced by, the possibility of formulating neuronal analogs of the different kinds of behavioral conditioning. From this perspective, it is useful to establish three dichotomies. First: learning can be **supervised** or **unsupervised**; in other words, with or without a teacher who makes explicit what has to be learned, either by fixing the order of presentation of the stimuli and showing the responses they have to trigger or by providing the proper reinforcements. Second: learning can take place in **open-loop** or in **closed-loop**, depending on whether the emitted responses have influence upon the subsequent stimulation. Third: learning can be **goal-directed** or **guided by a reinforcement signal**; in the first case, it proceeds by error correction, while, in the second, it consists of maximizing a function by trial-and-error.

There are two paradigmatic kinds of conditioning: classical (Pavlov, 1927) and instrumental (Skinner, 1938). The first, associative in nature, is elicited by the concomitance of two phenomena. The second takes place when applying certain reinforcement schemes, according to the spontaneous behavior displayed. Both lead to an increment (or decrement) in the frequency of appearance of the conditioned response, in the presence of the prefixed experimental conditions. In the dichotomic terms expressed above, classical conditioning takes place in open-loop and may or may not be goal-directed; instrumental conditioning, on the other hand, occurs necessarily in closed-loop and requires a reinforcement signal. Both may be supervised or unsupervised, since this depends on the experimental situation and not on the learning process itself. The different neuronal models of learning proposed postulate rules of modification of synaptic efficacy which, at the unineuronal or multineuronal level, constitute analogs of the two kinds of conditioning described. We must mention that when the analog is given at the unineuronal level, a network of such analogs can generate other types of behaviors and vice versa. Arbib et al. (1976), Sutton and Barto (1981) and Levine (1984) provide excellent systematizations and critical reviews of the different models proposed. There is a clear predominance of works relative to open-loop learning rules over those that involve closed-loop rules. Next, we describe five open-loop learning rules, three of them unsupervised (Hebb, Uttley and Barto-Sutton) and the remaining two supervised (perceptron, Widrow-Hoff); as well as two closed-loop rules, one unsupervised (Klopf) and the other supervised (associative search).

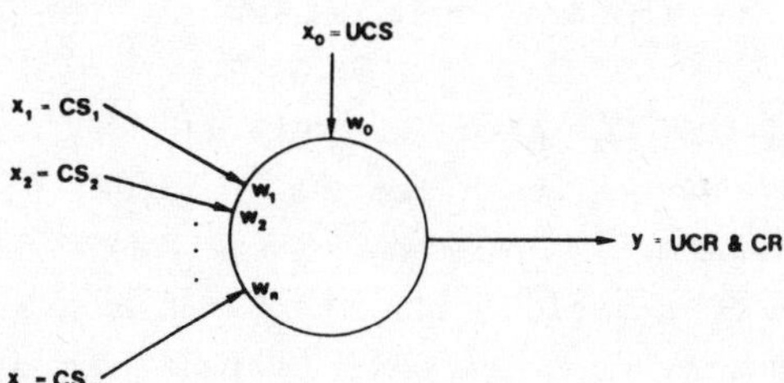

*** Figure 2.14 - Generic neuron model used as reference when describing the different learning rules proposed (from Sutton and Barto, 1981).

For the exposition of the most relevant learning rules formulated, we will use as reference the generic neuron model shown in Figure 2.14 and represented by the following expression:

$$y(t) = f\left[\sum_{i=1}^{n} w_i(t)\ x_i(t)\right] \qquad (2.4)$$

where $x_1, \ldots x_n$ are the inputs and $w_1, \ldots w_n$, their respective weights. The z that appears in Figure 2.14, but not in equation 2.4, is a specialized input that will be used by the teacher or the environment to provide information about the goal to be achieved or reinforcement signals and, consequently, will intervene in the formulation of the supervised rules.

2.2.2.1 - Neural analogs of classical conditioning

In classical conditioning, a neutral conditioned stimulus (CS) paired repeatedly with an unconditioned stimulus (US) that provokes an unconditioned reflex response (UR) finally elicits a conditioned response (CR), which completely or partly reproduces the UR.

The Hebbian learning rule, which implies an increment in the efficacy of the synapses that are active when the postsynaptic neuron fires, is the rule that has had the most powerful influence on the development of neuronal models of learning. Its expression in terms of the above generic model is:

$$w_i(t + 1) = w_i(t) + cx_i(t)y(t) \qquad (2.5)$$

where c is a positive constant that determines the speed of learning.

This rule does not use the specialized input z, so it constitutes an analog of open-loop unsupervised learning, guided neither by the objective nor by reinforcement. It is easy to see that this rule implements the interpretation of classical conditioning as association of stimuli by temporal contiguity: one has only to pair repeatedly an x_1=US with an x_2=CS, assigning initially to w_1 a value high enough to force the neuron to fire when activating x_1; after some pairings, the neuron will discharge when only x_2 is presented.

The strong influence of this rule results from how simply it explains the similarity between the UR and the CR, as well as the experimental fact that a first CS becomes US for a second CS.

The Hebbian scheme, further elaborated by Brindley (1969), has been incorporated into many networks proposed as models of different nervous structures: the cerebellum (Marr, 1969; Albus, 1971; Grossberg, 1972; Ito, 1982), the hippocampus (Marr, 1971; Kilmer and Olinski, 1974), the neocortex (Marr, 1970) and the retina (Grossberg, 1972). Since there are no detailed input/output electrophysiological specifications of these structures, the contribution of the mentioned models is more at the level of synthesis of networks capable of recognizing previously experienced stimuli, than at the level of analysis of the corresponding regions in the brain. Since the connections are reinforced by use, the response of the networks to a spatial pattern of stimulation increases with the relative frequency of presentation of this pattern. The above principle also explains the fact that each neuron of von der Malsburg's model (1973) becomes sensitive to a different orientation on the plane of a rectilinear stimulus.

A slightly different context of application is the formation of retinotopic connections in the visual cortex during development (Willshaw and von der Malsburg, 1976). The mathematical analysis of this model (Takeuchi and Amari, 1979; Amari, 1980; Kohonen, 1982) shows the restrictions that must be imposed on the initial connectivity of the network and on the propagation of excitation, to assure convergence towards a retinotopic connectivity.

Another use of the Hebbian rule, intimately related to the preceding one, though adhering to a rather different interpretation, is found in the implementation of associative memories (Spinelli, 1970; Nakano, 1972; Amari, 1977a & b; Anderson et al., 1977; Kohonen, 1977). The network shown in Figure 2.15 transforms the stimulation patterns $X=(x_1,\dots x_n)$ into the response patterns $Y=(y_1,\dots y_m)$. The inputs z_i act upon the neurons in the same way as the x_i, but are used to specify the patterns $Z=(z_1,\dots z_m)$ that have to be associated with the stimulation patterns X. The repeated presentation of k different stimulus pairs $(X_1,Z_1),\dots(X_k,Z_k)$, of which $X_1,\dots X_k$ form an orthogonal set, causes the network to learn to respond to the pattern Z_j when it receives the stimulus X_j, $j=1,\dots k$. The matrix of synaptic weights $[w_{lm}]$ converges approximately to the correlation matrix of the patterns X_j and Z_j, $j=1,\dots k$. For it to converge exactly to this matrix, a slight modification has to be introduced in the Hebbian rule.

The three most interesting aspects of this kind of network: resistance to noise, addressing by content, and generalization capability, derive from the distributed way in which information is stored. It seems from analytic studies that any application of spatial correlation can be structured in a way that permits its implementation through the Hebbian rule. There is a large probability that the converse is also true, in the sense that spatial correlation is the most complex function implementable through networks of Hebbian neurons, even in the case of incorporating into them delays or other modifications.

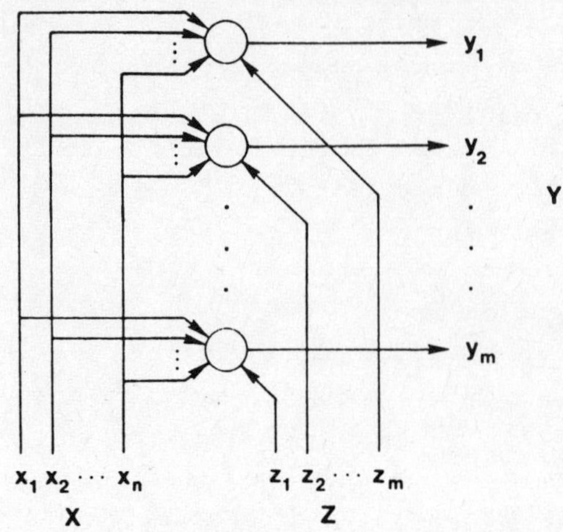

*** Figure 2.15 - Generic form of associative memories consisting of a set of m adaptive elements, which share the same m input channels (from Sutton and Barto, 1981).

The perceptron learning rule (Rosenblatt, 1962) had traditionally been considered an analog of instrumental conditioning (Arbib et al., 1976). Recently, though, Sutton and Barto (1981) have shown that it can be reinterpreted as an open-loop rule, guided by a pre-established objective, thus becoming an analog of classical conditioning. Its expression in terms of the generic model that we use as reference (equation 2.4) is:

$$w_i(t+1) = w_i(t) + c(z(t)-y(t))\, x_i(t) \qquad (2.6)$$

where y(t) and z(t) are binary and, in the model, f is a threshold function.

The perceptron --written in lower case as suggested by Rosenblatt, who wanted to emphasize in this way that it is a class of networks and not a specific network-- works essentially as a classifier of stimulation patterns. Its simplest version, represented in Figure 2.16, has three layers: sensory, preprocessor, and actuator. Only the weights of the connections between the second and the third layer can be modified. If z(t) supplies the desired classification for each pattern $X=(x_1,\ldots x_n)$

and the classes are linearly separable, then the synaptic weights will converge to a configuration that classifies all patterns correctly. The rule then provides an iterative algorithm to find the solution of a system of linear inequalities.

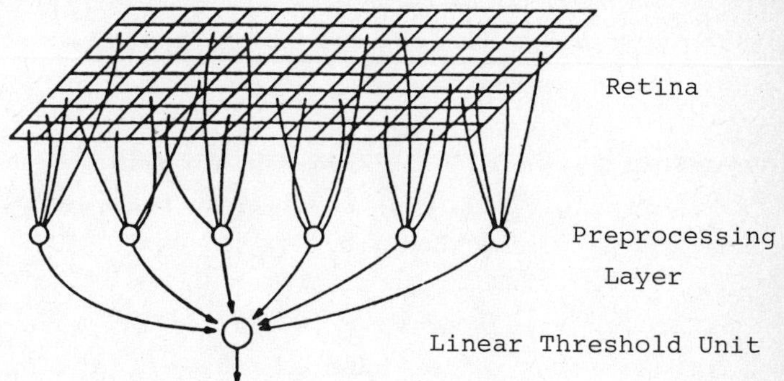

*** Figure 2.16 - Perceptron with preprocessing layer (from Arbib et al., 1976).

Rosenblatt generalized this simple version of perceptron in different ways: incorporating more actuators, considering a variable number of intermediate layers, and introducing both intra- and inter-layer feedback. Nilsson (1965) proved convergence theorems for different types of perceptron, always presupposing the existence of a solution, or in other words, of a configuration of weights that would give rise to the correct classification. Minsky and Papert (1969) approached the complementary question, characterizing the limits of the discriminative ability for each type of perceptron. Essentially, what they did was to fix the number of layers and catalog the classification problems according to the number or the dispersion of inputs to each neuron necessary to solve them. Winograd (1965,1967), Spira and Arbib (1967) and Spira (1969) studied the same question, but from a dual point of view: they fixed the number of inputs per neuron and determined the minimum number of layers required to solve different problems.

The fact that the perceptron rule has traditionally been conceptualized as an analog of instrumental conditioning derives from considering the term (z(y)-y(t)), in equation 2.6, as a reinforcement signal provided by the environment. This is the way it worked in Rosenblatt's original description, where the teacher supplied an indication of the direction in which synaptic weights had to be modified, depending on which the output had been. Nevertheless, Sutton and Barto (1981) point out that the above rule is a particularization to the binary case of the more generic Widrow-Hoff rule which, as we will next see, coincides with the Rescorla-Wagner model of classical conditioning. In the latter rule, the term (z(t)-y(t)) becomes clearly an error signal between the predicted classification y(t) and the desired one z(t) which, being so stereotyped, can be calculated by the system itself, so that the environment simply provides the correct classification through the specialized input.

The Widrow-Hoff learning rule (1960) is expressed by the same equation 2.6, but considering $x_i(t)$, y(t) and z(t) to take real values and the f of equation 2.4 to be the identity.

Through this rule, the repeated presentation of pairs (X_i, z_j), j=1,...k, with $X_1,...X_n$ linearly independent, causes the convergence of the synaptic weights toward the proper configuration for the response to each stimulus X_j to be the desired real number z_j. The rule thus provides an iterative procedure to find the solution of a system of linear equations. If the patterns $X_1,...X_k$ are not linearly independent, the rule can be slightly modified (converting the parameter c into a variable that goes to zero with time) so that the convergence of the weights minimizes the quadratic error between the actual output and the desired one. This slightly modified version of the rule computes a linear regression iteratively (Duda and Hart, 1973).

Kohonen and Oja (1976), Amari (1977a & b) and Albus (1979) have studied the implementation of associative memories through networks of neurons that incorporate the Widrow-Hoff rule. While, as we saw previously, the Hebbian rule only provides a perfect association if the stimulation patterns are orthogonal, the Widrow-Hoff rule relaxes the restriction to their being linearly independent. As could be foreseen, if c approaches zero with time, optimal associations are formed according to the criterion of quadratic minimization, even when the stimulation patterns are linearly dependent.

The Widrow-Hoff rule was enunciated in the technological context of adaptive electronic circuits. Independently, Rescorla and Wagner (1972) proposed their theory of classical conditioning, which describes a great variety of effects observed in animal learning experiments and is expressed by an equation identical to 2.6, with the same restrictions imposed by Widrow-Hoff. Such a remarkable coincidence --first noted by Sutton and Barto (1981)-- has renewed the interest in this rule, of which several variants have been proposed, among which we will point out its predictive version (Tsukahara and Kawato, 1982):

$$w_i(t+1) = w_i(t) + c(z(t)-y(t-\Delta t))\ x\ (t-\Delta t) \qquad (2.7)$$

and the rules by Uttley and by Barto-Sutton, that we will next describe.

Uttley's learning rule (1970) is a hybrid between the Widrow-Hoff and the Hebbian rules and its representation can be made to approach one or the other:

$$\begin{aligned} w_i(t+1) &= w_i(t) + c(z(t)-y(t))\ x_i(t) = \\ &= w_i(t) + c\left[-w_0x_0 - \sum_{i=1}^{n} w_i(t)x_i(t)\right]x_i(t) = \\ &= w_i(t) + c\left[-\sum_{i=0}^{n} w_i(t)x_i(t)\right]x_i(t) = \\ &= w_i(t) - cy_0(t)\ x_i(t) \end{aligned} \qquad (2.8)$$

where $z(t)=-w_0x_0(t)$ and the f of equation 2.4 is the identity.

The inclusion of the minus sign, that differentiates it from the Hebbian rule, gives it greater stability, since equilibrium is attained when the addition of the inputs multiplied by their respective weights is zero. Also, as Uttley (1975) himself pointed out, this rule reproduces a large part of the effects taken into consideration by the Rescorla-Wagner model, without needing to incorporate any specialized input. Conversely, it has the disadvantage of requiring that the unconditioned stimulus inhibits the conditioned response, a hypothesis that is completely unsustainable at the behavioral level.

The Barto-Sutton learning rule (1982) is expressed by the following equation:

$$w_i(t+1) = w_i(t) + c\left[y(t)-\overline{y}(t)\right]\overline{x}_i(t) \qquad (2.9)$$

where all variables are real, the f of equation 2.4 is the identity, and $\overline{x}_i(t)$ and $\overline{y}(t)$ are weighted averages of the values of $x_i(t)$ and $y(t)$ during a time interval preceding T and finishing at this instant.

By means of this rule, the neuron model works as a predictor of its own subsequent response. Analogous to the Hebbian and Uttley's rules, the present one does not require any specialized input and, differing from the predictive version of the Widrow-Hoff rule only in the substitution of z(t) by y(t) and in the use of the averages $\bar{x}_i(t)$ and $\bar{y}(t)$, it retains all the advantages of the latter rule, though presupposing the existence of intraneuronal mechanisms that calculate the temporal average of each input and of the output.

Networks of neurons equipped with this rule exhibit the same properties as isolated neurons, in addition to storing the information in an associative and distributed form.

2.2.2.2 - Neural analogs of instrumental conditioning

Instrumental conditioning requires the generation of spontaneous behaviors by the individual, which could be later favored or weakened through the application of certain reinforcement schemes. Conceptually, it is similar to the Darwinian process of mutation and natural selection, to learning by trial-and-error and to the generation-and-test strategies used in Artificial Intelligence.

Very little work has been done on neuronal analogs of instrumental conditioning, probably because of the difficulty of determining which signals are susceptible to acting as reinforcement at the level of a single neuron. Still, the possibility that the components of a system learn in closed-loop, without requiring the explicit presentation of the objective to be achieved, but only a measure of the adequacy of the emitted responses, is highly desirable if the purpose is to explain and reproduce the observed reality that most learning occurs in barely-informative environments.

All the proposed rules use random search procedures to maximize a reinforcement function (reward); thus they have their origin in the works by Tsetlin (1973), afterward systematized by Narendra and Thathachar (1974). Precedents worth mentioning, enunciated in areas slightly different from the neuronal one, are the adaptation strategy proposed by Widrow et al. (1973) and the ALOPEX system of Harth and Tzanakou (1974).

The heterostatic learning rule (Klopf, 1972) tries to maximize the level of depolarization of the neuron, modifying the synaptic weights according to the following expression:

$$w_i(t+1) = \begin{cases} w_i(t)+c\Delta u\ y(t-1)\ \bar{x}_i(t-1), & \text{if } w_i(t)\cdot\Delta u>0 \\ w_i(t), & \text{otherwise} \end{cases} \tag{2.10}$$

where $\Delta u=\sum_{j=1}^{n} w_j(t)\left[x_j(t)-x_j(t-1)\right]$, $\bar{x}_i(t)$ has the same meaning of weighted temporal average described before and the f of equation 2.4 is a binary threshold function.

The excitatory stimulation that arrives subsequent to the discharge of the neuron is interpreted as a positive reinforcement, and so it leads to an increment in the efficacy of the excitatory synapses that were active just before the discharge. Analogously, an increment in the inhibition after the firing of the neuron is considered an aversive reinforcement and determines an increment in the efficacy of the inhibitory synapses that were active previous to the discharge.

This rule has implicit the so-called "secondary reinforcement", so that any context previously positively reinforced becomes itself a positive reinforcement. Another interesting consequence is that the reinforcement signal does not require a specialized channel, but arrives through the synaptic inputs of the neuron.

The implications of incorporating this rule into the neurons of a network have not been studied as much as those derived from rules that use instead a specialized input channel for reinforcement, as we will next see.

The associative search learning rule (Barto et al., 1981) is expressed by the following equation:

$$w_i(t+1) = w_i(t)+c\left[z(t)-z(t-1)\right]y(t-1)\bar{x}_i(t-1) \tag{2.11}$$

where $\bar{x}_i(t)$ and z(t) are real, y(t) is binary and the f that characterizes the reference model (equation 2.4) is:

$$y(t) = f(\sum_{i=1}^{n} w_i(t)x_i(t)+s(t)) = \begin{cases} 1, \text{ if } \sum_{i=1}^{n} w_i(t)x_i(t)+s(t)>0 \\ 0, \text{ otherwise} \end{cases} \tag{2.12}$$

s(t) being a Gaussian stochastic process with zero mean.

A neuron model equipped with this rule learns to maximize z(t) for each stimulus situation. If z(t) is a random variable, its mathematical expectation is instead maximized. It is interesting to note that the procedure followed conforms to the trial-and-error one, since the response contains a random component.

The neural networks of the kind shown in Figure 2.15 that incorporate the above rule are called Associative Search Networks (ASN) and, if certain conditions are satisfied, they learn to respond to each stimulus situation $X_i=(x_{i1},\ldots x_{in})$ of a set $\{X_1,\ldots X_k\}$ repeatedly presented, with the vector $Y=(y_1,\ldots y_m)$ that maximizes the reinforcement function z. The conditions that have to be satisfied are: (a) the function z has to be unimodal, and (b) for each neuron, the subset of stimulus situations in which the optimum response is 0 has to be linearly separable from the corresponding subset in which the optimum response is 1.

To solve the problem that the transitions between stimulus situations could cause changes in the reinforcement function and, consequently, the difference (z(t)-z(t-1)) would lose informative value, Barto et al. (1981) incorporated into the network a predictor neuron that works according to the rule expressed by equation 2.9. To the advantages already described for classical associative memories --resistance to noise, generalization capacity, and addressing by content-- the above network adds two more: not needing the explicit presentation of the desired response for each stimulus situation, and using a single reinforcement signal common to all neurons.

The same kind of network has been successfully applied to the resolution of problems not approachable by means of the previously proposed rules, such as that of spatial orientation (Barto and Sutton, 1981a, 1983) and that of controlling an inverted pendulum (Barto et al., 1983).

2.3 - SUMMARY

In the present chapter we have outlined the reference framework for the work described in the following chapters. The exposition has been structured around the two poles expressed in the title: experimental data and previous models. In the first part, we have described the phenomena to be modelled, distinguishing between those relative to a single neuron and those involving a network. The second part has been devoted to the specification and systematization of the different neuronal models of autorhythmicity and of learning proposed by other authors, as well as to indicate their successes and shortcomings, and the types of questions they do and do not address.

Only modelling at the electrophysiological level has been considered; hence, other types of data and of models (biochemical, molecular, etc.) have only been dealt with marginally, when required by the exposition. The experimental description has been primarily about aspects linked to the response to, and learning of, rhythmic stimuli. This criterion has been maintained in the exposition of the biophysical models of pacemaker neurons, but not in dealing with the neuronal models of learning. The reason for this lack of homogeneity is the almost complete focussing of the latter research area in the learning of spatial patterns of stimulation. Two further characteristics that both define and limit this area are: the use of logical neuron models and the exclusive consideration of learning as changes in synaptic efficacy. The current trend is to shift the emphasis initially placed in the macroscopic emergence of computational capabilities from very simple elements (neurons), toward the incorporation of more and more of these capabilities into individual neurons.

The neuron and network models we will propose in Chapters 3 and 5, respectively, address the question of the learning of rhythmic stimuli from the double theoretical context described above, since, being based on a biophysical model of pacemaker neurons, they postulate a learning rule in which the modification takes place at the neuron soma and affects the **temporal** evolution of neuronal activity, thus constituting an alternative to the synaptic rules, which imply **structural** changes.

CHAPTER 3

MODELLING AND SIMULATION OF A PLASTIC PACEMAKER NEURON

> *"A theory has only the alternatives of being right or wrong. A model has a third possibility: it might be right but irrelevant".*
>
> *M. Eigen*

The neuron model proposed in this chapter is intended to respond to two very different kinds of requirements that will be described in the first section. On the one hand, it must agree with the experimental evidence at the unineuronal level described in the preceding chapter. On the other hand, it has to constitute an adequate basic node with which to set up a network able to reproduce the experimental results at the multineuronal level.

The second section contains the description of our nonlinear continuous-time model. It differs from similar ones, like those proposed by Perkel et al. (1964), Junge and Moore (1966), Walloe et al. (1969), McGregor and Oliver (1974) and Hartline (1976b), in the way of introducing randomness, of determining the output amplitude and, above all, in the incorporation of a certain learning capacity.

In the proposed model, randomness derives from fluctuations in the equilibrium level of the membrane potential, which is sampled from a Gaussian distribution after each discharge of the neuron.

The output amplitude depends on the difference between the magnitudes of the postsynaptic potential and the threshold, at the time of firing.

In relation to learning, we propose an alternative to Hebbian-like rules which, despite their wide use in the theoretical field (Rosenblatt, 1962; Brindley, 1969; Minsky and Papert, 1969; Nakano, 1972; von der Malsburg, 1973; Wigström 1973; Grossberg, 1974; Amari, 1977b; Anderson et al., 1977; Kohonen, 1977; Uttley, 1979), have not obtained much experimental support (von Baumgarten, 1970; Kandel, 1976, 1978; Pinsker and Kandel, 1977; Sutton and Barto, 1981; Alkon et al., 1982). The idea underlying the mentioned alternative is to substitute structural changes in the efficacy of synaptic connections for changes in the temporal firing pattern of neurons. This general principle has been realized in different ways, giving rise to several learning rules, the behavior of which has been studied through simulation and contrasted with experimental evidence of the assimilation of rhythms by single pacemaker neurons.

The simulation results, included in the third section, show the agreement between the functioning of the model and the experimental data, in the situation previous to learning, and the superiority of one of the learning rules over the others in reproducing certain phenomena relative to plasticity in the firing pattern. Some of these results had previously been presented and debated in Torras (1981, 1982a & b).

3.1 - THEORETICAL AND EXPERIMENTAL REQUIREMENTS

One of the objectives pursued with the neuron model proposed is, as said in the introductory chapter, to obtain a basic node for a network that permits analyzing the microscopic substratum of the macroscopic behavior shown by certain neuronal structures, when submitted to the characteristic experimental conditions of the tracer technique (Section 2.1.2), with the intention of proposing an explanation at the microscopic level of that behavior. This technique imposes the following theoretical requirements on the neuron model:

(a) To be specified in terms of electrophysiological variables, the temporal trajectory of which can be easily followed and recorded.

(b) To be equipped with the capability of learning temporal aspects of the stimulation.

(c) To include pacemaker neurons as a particular case, since they are the nervous cells in which temporal aspects are more obvious.

With regard to the experimental requirements, we will require that the model's behavior, in its pacemaker version, reproduces the following phenomena (Section 2.1.1):

(a) Spontaneous interspike interval variability, according to a certain probability distribution.

(b) Phase-shifts provoked by occasional perturbations.

(c) One-to-one, subharmonic and superharmonic entrainment.

(d) Long-lasting changes in the firing pattern (plasticity).

3.2 - THE PACEMAKER NEURON MODEL PROPOSED

We have taken Perkel et al.'s model as our starting point, because, among all those specified in terms of electrophysiological variables, it is the one that has the level of complexity and the types of relationships between variables most adequate to our needs. In fact, the model satisfies a great part of the requirements listed in the preceding section --it includes pacemaker neurons as a particular case and reproduces phenomena such as phase-shifts and entrainment-- and its variables and the relationships between them are explicited at a level of detail that makes it possible to incorporate aspects, such as randomness and a learning capacity, that lead to the satisfaction of the remaining requirements.

3.2.1 - Informal description

The two variables that characterize the short-term state of the neuron are: the membrane potential (P) and the threshold (H). The membrane potential has a spontaneous component (Pb) and a postsynaptic component (Ps): P=Pb+Ps.

Between discharges, H and Pb follow opposite exponential trajectories, the first one decreasing and the second one increasing, with time-constants τ_h and τ_b, respectively.

The postsynaptic component reflects the incidence of the external stimulation (x) upon the membrane potential. Ps accumulates the excitatory inputs positively and the inhibitory ones negatively, as they arrive, so that spatial summation is thus incorporated into the model. Temporal summation, on the other hand, is made possible by the exponential maintenance of Ps, which tends toward zero with a time-constant τ_s.

In synthesis, while P remains lower than H, the model behaves in a linear way and the dynamics of the variables defined up to now are governed by the following differential equations:

$$\frac{dPb}{dt} = - \tau_b \cdot (Pb - Pb_1)$$

$$\frac{dPs}{dt} = - \tau_s \cdot Ps + x \qquad (3.1)$$

$$\frac{dH}{dt} = - \tau_h \cdot (H - H_1)$$

where Pb_1 and H_1 are the limit values toward which Pb and H tend asymptotically, and x is the sum of input impulses ariving at the different synapses.

When P grows to the point of surpassing H, the neuron fires and the three variables Pb, Ps and H are reset to the predefined constant values Pb_0, 0 and H_0, respectively. This is the point at which the nonlinearity arises and the discontinuous way in which it happens shows the convenience of defining another variable T, that records the time elapsed since the last discharge.

Immediately after the discharge, the neuron enters an absolute refractory period, in which the stimulation has no effect, followed by a relative refractory period, when only a very intense input can make the neuron fire, since the spontaneous potential and the threshold are just beginning to follow their opposed exponential trajectories.

If a value larger than H_1 is assigned to Pb_1, the neuron behaves as an oscillator. Figure 3.1 contains a graphical representation of the evolution of the state-variables of the model in this case.

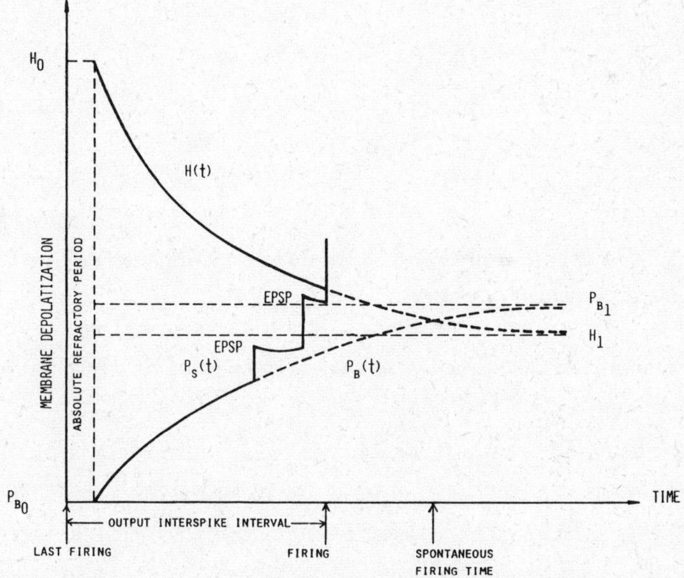

*** Figure 3.1 - Dynamics of the pacemaker neuron model: Membrane depolarization. See text for details.

The first innovation of the proposed model with respect to Perkel et al.'s is to consider the parameter Pb_1 as a stochastic process, of constant value in the interspike interval and of Gaussian probability density with mean m_{Pb_1} and variance σ_{Pb_1} at the time of firing.

The second innovation is to consider the output amplitude to be an increasing function of the difference between the membrane potential and the threshold at the time of firing (this difference occurs because input impulses add instantaneously to the membrane potential). The output is minimum when the neuron fires spontaneously and increases both with the amplitude of the stimulation that triggers the discharge and with the delay in the arrival of this stimulation.

Finally, the fundamental innovation is to consider the parameter m_{Pb_1}, the mean value of Pb_1 at the time of firing, to be a variable, subject to a certain learning rule that equips the model with the desired adaptation capacity. The idea underlying all the learning rules tested consists in increasing the value of m_{Pb_1} in certain cases in which the neuron is forced to fire, and decreasing it in others in which the stimulation arrives after the discharge. By means of this mechanism, the neuron tends to accommodate its spontaneous firing rate to the one imposed by the stimulation. Thus, learning introduces a new nonlinearity into the model, that also takes place at the time of firing.

The three learning rules studied can be expressed in a general form as follows:

$$m_{Pb_1}^{+} = F_c(m_{Pb_1}^{-}, Ps, T) \tag{3.2}$$

where $m_{Pb_1}^{-}$ and $m_{Pb_1}^{+}$ are the values of the parameter m_{Pb_1} before and after firing, and c is a constant that determines the rate of modification of the mentioned parameter.

The first learning rule tested follows a goal-driven strategy, since it takes into consideration the ideal value m_{Pb_1} the asymptotic limit Pb_1 should have, so that, without stimulation, the neuron would fire at a frequency of period equal to the last interspike interval T_s. This ideal value can be calculated through the following expression:

$$\mathfrak{I}m_{Pb_1} = \frac{-Pb_0 \cdot e^{-\tau_b \cdot T_s} + (H_0 - H_1) \cdot e^{-\tau_h \cdot T_s}}{1 - e^{-\tau_h \cdot T_s}}$$

The strategy is to make the value of the parameter m_{Pb_1} approach this ideal, either adding to it or subtracting from it the product of c by the difference between both, as indicated in the equation:

$$m_{Pb_1} = \begin{cases} m_{Pb_1} + c(\min(\overline{m_{Pb_1}}, Jm_{Pb_1}) - m_{Pb_1}), \text{ if } Jm_{Pb_1} \geqslant m_{Pb_1} \\ \\ m_{Pb_1} + c(\max(\underline{m_{Pb_1}}, J_{m_{Pb_1}}) - m_{Pb_1}), \text{ otherwise} \end{cases} \tag{3.3}$$

where $\overline{m_{Pb_1}}$ and $\underline{m_{Pb_1}}$ are the maximum and minimum values that m_{Pb_1} can adopt.

The second learning rule follows a strategy dependent on the size of the postsynaptic potential at the time of firing. The rule will only have effect when this potential is positive and, depending on whether it surpasses a specific critical value $P^{*}>0$, the value of the parameter m_{Pb_1} will increase or decrease by a constant quantity c:

$$m_{Pb_1} = \begin{cases} \min(m_{Pb_1}+c, \overline{m_{Pb_1}}), \text{ if } Ps \geqslant P^* \\ \\ \max(m_{Pb_1}-c, \underline{m_{Pb_1}}), \text{ if } P^* > Ps \geqslant 0 \\ \\ m_{Pb_1}, \text{ otherwise} \end{cases} \tag{3.4}$$

The third learning rule is obtained from the second by modifying its applicability conditions. In it, decremental learning only takes place when an excitatory stimulus arrives just after the discharge; that is to say, while T belongs to the interval $(0,T^{*}]$, T^{*} being a parameter to be determined. Although, in this rule, learning can take place when there is no discharge, its effect upon the behavior of the model will not be noticed until a new value for Pb_1 is generated at the next time of firing. In order to undertake the formalization of the model, which we will do in Section 3.2.3, it is advisable to separate the rule into two equations:

$$m_{Pb_1} = \begin{cases} \min(m_{Pb_1}+c, \overline{m_{Pb_1}}), \text{ if } Ps \geqslant P^* \\ \\ m_{Pb_1}, \text{ otherwise} \end{cases} \tag{3.5}$$

$$m_{Pb_1} = \begin{cases} \max(m_{Pb_1} - c,\ \underline{m_{Pb}}_1), & \text{if } T \leqslant T^* \\ m_{Pb_1}, & \text{otherwise} \end{cases} \qquad (3.6)$$

Equation 3.5 will be applied at the time of firing and equation 3.6, at the time of arrival of an excitatory stimulus.

We shall report on the evaluation of those three rules in Section 3.3.4.

3.2.2 - A priori neurophysiological validation of the model

The objective of this section is to provide experimental evidence to support the three innovations the proposed model introduces. The validation will be done at the level of mechanisms, leaving the behavioral verification for the next section.

Many experimental studies indicate that the fluctuations observed in the duration of the interspike interval of a pacemaker neuron are due to changes in the temporal evolution of the membrane potential (Junge and Moore, 1966; Calvin, 1974; Bustamante et al., 1981). However, and because of the fact widely reflected in the literature (Bullock, 1976; Bustamante et al., 1981) that the firing threshold does not remain constant, many stochastic neuron models have incorporated the source of randomness into the threshold (Moore et al., 1966; Walloe et al., 1969; Knight, 1974). The proposed model takes into account both aspects of the experimental evidence: On the one hand, it makes the time of firing dependent on the trajectory followed by the spontaneous membrane potential, which varies from one interval to the next; on the other hand, because of the fixed evolution of the threshold in each interval, its value at the time of firing varies from time to time.

Concerning the second innovation, the experimental data cover almost the entire spectrum of possibilities. There are pacemaker neurons with a practically inalterable amplitude and frequency of firing; others codify the amplitude of the stimulation received in terms of frequency, leaving the amplitude unchanged; some act as nonlinear transducers of the input intensity; finally, the ones that have a certain plastic capacity, reflected in phenomena such as habituation, sensitization, facilitation, adaptation to imposed rhythms, etc. usually vary the amplitude of their response according to previous experience (Parnas et al.,1974; Bullock, 1976; Ayers and Selverston, 1977b; Gillette et al., 1980; Holden and Ramadan, 1981b; Pinsker and Ayers, 1983). Since our interest is primarily centered upon plastic phenomena, it seems proper to have in the model the possibility of varying the output amplitude, according to the stimulation conditions. Because of the way this possibility has been realized, output increases as the learning of the rhythm imposed by the stimulation progresses.

While the existence of a certain plasticity in the firing pattern of some pacemaker neurons seems firmly established, the mechanisms underlying this capacity are, on the contrary, the object of serious controversy. The most notable landmark relative to this question is the evidence given by Pinsker and Kandel (1977) that the mechanism explaining the plasticity shown by certain pacemaker neurons of the abdominal ganglion of Aplysia in their firing pattern does not reside in the synapse, as Hebbian-like rules postulate, but depends on the intrinsic oscillatory properties of the postsynaptic neuron. The data contributed by these two authors give empiric support to the initial suggestion by Strumwasser (1965, 1967), and subscribed to afterwards by Frazier et al. (1965), von Baumgarten (1970), Parnas et al. (1974) and Kandel himself (1976), that the mechanism of generation of the endogenous rhythms could be the substratum of certain plastic phenomena.

The three learning rules set up here respond to this evidence, incorporating plasticity in one of the parameters that determine the trajectory followed by the membrane potential in its spontaneous growth, which, in the final analysis, determines the firing rate of the neuron.

Furthermore, von Baumgarten (1970) observed, again in pacemaker neurons of Aplysia, that the slope of the mentioned spontaneous potential increased when the neuron was submitted to stimulation frequencies higher than the spontaneous one and decreased when frequencies lower than the spontaneous one were applied to it. These are exactly the effects of accelerative and decelerative learning, respectively, as they have been implemented in the three rules described: An increment (decrement) in m_{Pb_1} leads to an increase (decrease) in the slope of the spontaneous potential.

The differences between the rules reside in the conditions in which each kind of learning takes place. For the model to be biologically plausible, these conditions must be expressed in terms of the instantaneous state of the neuron. **The first rule** takes into consideration an ideal value that has to be computed from the duration of the last interspike interval and so requires an additional variable that counts the time elapsed between discharges, as well as a high computational capability.

The second rule seems more plausible, since it is based on only one variable --the postsynaptic potential-- that characterizes the instantaneous state of the neuron. Nevertheless, the need for a critical value that marks the frontier between the accelerative and decelerative types of learning is more difficult to justify.

Finally, **the third rule** presupposes that the discharges and the excitatory stimuli leave different kinds of ephemeral traces in the state of the neuron; the arrival of an excitatory stimulus while the trace left by a discharge remains elicits decelerative learning and, conversely, the generation of an action potential while the trace left by an excitatory stimulus remains triggers the acceleratory learning process. Possible biochemical substrata of traces of this kind have been conjectured by Sutton and Barto (1981) and Barto and Sutton (1982).

Since the model, with the different learning mechanisms, has been described in electrophysiological terms, the validation has been carried out at this level. The formulation of possible biochemical realizations of these mechanisms --changes in the permeability of the membrane to the different ions, metabolic processes that activate electrogenic pumps, synthesis and transport of proteins and enzymes, etc.-- are far from the objectives of the present work.

3.2.3 - Formal specification

Zeigler (1976) establishes six dichotomies that permit characterizing and classifying models, and proposes a methodology to unify their formal presentation. Both initiatives attempt to structure the modelling domain and are especially useful when, as now, we try to locate a model.

We next list the six above-mentioned dichotomies, highlighting the poles that characterize the model proposed in this thesis:

Continuous-time / Discrete-time

Continuous-state / Discrete-state

Deterministic / **Stochastic**

Autonomous / **Non-autonomous**

Time-invariant / Time-variant

Adaptive / Non-adaptive

Concerning the formalization, the model shows the difficulty of being partially specified in terms of differential equations and partially in terms of discrete changes of state when certain conditions are met. It does not seem appropriate, then, to use a Differential Equation System Specification (DESS). On the other hand, a Discrete EVent system Specification (DEVS) is not proper either, because the arrival of an input impulse can occur at any moment. Since all the differential equations that appear in the model are easy to integrate and the formalization is eventually aimed at its simulation, which will be carried out on a digital computer, it seems convenient to discretize the time and use a Discrete Time System Specification (DTSS), that corresponds to a synchronous sequential machine (but with continuous input, output and state-variables).

The discrete-time neuron model will then be represented by the quintuple $(X, S, Y, \delta, \lambda)$, where

$X = \mathbb{R}$ is the set of inputs

$S = [-1, \infty) \times \mathbb{R} \times \mathbb{R} \times [\underline{m_{Pb_1}}, \overline{m_{Pb_1}}]$ is the set of states

$Y = [0, M_o]$ is the set of outputs

$\delta_h : S \times X \longrightarrow S$ is the transition function

$$\delta_h(T, Ps, Pb_1, m_{Pb_1}, x) = \begin{cases} (T+h,\ Ps.e^{-\tau_s \cdot h}+x,\ Pb_1,\ \overset{(1)}{F_c}(m_{Pb_1}, Ps, T)) \\ \quad \text{if } T \neq -1,\ P < H \wedge x > 0 \\ (T+h,\ Ps.e^{-\tau_s \cdot h}+x,\ Pb_1,\ m_{Pb_1}) \\ \quad \text{if } T \neq -1,\ P < H \wedge x \leqslant 0 \\ (-1,\ Ps,\ Pb_1,\ \overset{(2)}{F_c}(m_{Pb_1}, Ps, T)) \\ \quad \text{if } T \neq -1 \wedge P \geqslant H \\ (0,\ 0,\ Z,\ m_{Pb_1}),\ \text{if } T=-1 \end{cases} \qquad (3.7)$$

$\lambda : S \longrightarrow Y$ is the output function

$$\lambda(T, Ps, Pb_1, m_{Pb_1}) = \begin{cases} M_o - R_o \cdot e^{(H-P)}, & \text{if} \quad P \geqslant H \\ 0, \text{otherwise} \end{cases}$$

with:

$$P = Pb_1 + (Pb_0 - Pb_1) \cdot e^{-\tau_b \cdot T} + Ps$$

$$H = H_1 + (H_0 - H_1) \cdot e^{-\tau_h \cdot T}$$

and m_{Pb_1}, m_{Pb_1}, M_o, R_o, τ_s, Pb_0, τ_b, H_1, H_0 and τ_h being parameters of the model, h the time increment, and Z a random variable of Gaussian probability density, $N(m_{Pb_1}, \sigma_{Pb_1})$.

The learning functions $F_c^{(1)}$ and $F_c^{(2)}$ are of the kind described in Section 3.2.1. If the first or the second learning rules are implemented, $F_c^{(1)}$ has to be substituted by the identity with respect to m_{Pb_1} and $F_c^{(2)}$ is given by equation 3.3 or 3.4, depending on whether it is one case or the other. If the third learning rule is implemented, $F_c^{(1)}$ and $F_c^{(2)}$ are given by equations 3.6 and 3.5, respectively. It is important to mention that only in the latter case is it necessary to specify separately the first two branches of equation 3.7.

Table 3.1 aims to clarify the neurophysiological meaning of the variables and constants that appear in the above formal description.

VARIABLES

x : Stimulation amplitude.

T : Time elapsed since the last discharge.

Ps : Postsynaptic potential.

Pb_1 : Asymptotic limit of the spontaneous potential.

m_{Pb_1}: Mean value of the Gaussian distribution of the asymptotic limit of the spontaneous potential.

Pb : Spontaneous membrane potential.

P : Membrane potential.

H : Threshold.

y : Firing amplitude.

CONSTANTS

Pb_0 : Value of the membrane potential after discharge.

$\underline{m_{Pb_1}}$: Minimum mean value of the asymptotic limit of the spontaneous membrane potential.

$\overline{m_{Pb_1}}$: Maximum mean value of the asymptotic limit of the spontaneous membrane potential.

$\sigma^2_{Pb_1}$: Variance of the asymptotic limit of the spontaneous membrane potential.

τ_b : Time-constant of the spontaneous membrane potential.

τ_s : Time-constant of the postsynaptic potential.

H_0 : Value of the threshold after discharge.

H_1 : Asymptotic limit of the threshold.

τ_h : Time-constant of the threshold.

M_o : Maximum output amplitude.

R_o : Range of variation of the output amplitude.

c : Learning rate.

P^* : Critical value of the postsynaptic potential, above which the accelerative learning acts in the second and third rules.

T^* : Critical value of the time elapsed since the last discharge, above which the decelerative learning does not act in the third rule.

*** Table 3.1 - List of variables and constants of the neuron model.

3.3 - SIMULATION AND A POSTERIORI VALIDATION OF THE MODEL

"No one would suppose that we could produce milk and sugar by running a computer simulation of the formal sequencies in lactation and photosynthesis".

J. Searle, 1980

The discrete-time model described in the preceding section was implemented first on a VAX-11/780 (Department of Computer and Information Sciences, University of Massachusetts at Amherst) and afterwards on a SEL 32/77 (Institut de Cibernètica, Universitat Politècnica de Catalunya), together with a set of graphical and signal processing routines that permit representing on a Grinnell screen or on a printer the trajectories followed by the different variables through time, as well as histograms of interspike intervals, autocorrelation functions, discrete Fourier transforms, peristimulus histograms, etc.

In the present chapter, we do not attempt to reproduce quantitatively the behavior of a specific neuron, but obtain a model that, by its potential to generate the phenomena observed in the behavior of pacemaker neurons, constitutes a good basic node for the network. Given the qualitative character of the objectives sought, it seemed unnecessary to adjust the parameters with respect to a specific neuron and statistical criteria have been followed to fix their values on the basis of the experimental data reported in the literature.

To the parameters of the proposed model already appearing in the previous model by Perkel et al. (1964), we have assigned the values reported by these authors, which were fixed in agreement with the values observed in pacemaker neurons of the abdominal ganglion of Aplysia. They are the following: Pb_0=-100 mv, τ_b=6.93 s^{-1}, τ_s=6.93 s^{-1}, H_0=-20 mv, H_1=-50 mv, τ_h=3.43 s^{-1}. In addition, Perkel et al. assigned a constant value of -40 mv. to Pb_1, which fixes the spontaneous firing period at 430 ms and the frequency at 2.3 Hz. In the proposed model, Pb_1 is variable, the superior and inferior limits of its mean value having been fixed at m_{Pb_1}=-33 mv and m_{Pb_1}=-46 mv, respectively, to cover a range of spontaneous periods between 325 and 630 ms and a band of frequencies between 1.6 and 3 Hz.

Different values --always congruent with the experimental data brought by Junge and Moore (1966)-- have been assigned to σ_{Pb_1}, to analyze the effect of the interspike interval variability upon the patterns of entrainment between the neuron and different stimulation frequencies.

To obviate the question of whether facilitation takes place pre- or post-synaptically (Parnas et al., 1974), and since our interest is in the propagation of influences in the network, the output has been interpreted in terms of its effect upon the membrane potential of the postsynaptic neuron. According to the data supplied by Parnas et al. (1974), the values M_o=30 mv and R_o=25 mv have been assigned.

The values of the parameters c, P^* and T^*, that characterize functionally the learning rules, have been adjusted so that the behavior of the model reproduces the desired experimental phenomena. The observation that this behavior changes qualitatively when the values of those parameters vary beyond a very limited range, makes the prediction about the underlying mechanisms much more restrictive, and, consequently, easier to refute.

Finally, we mention that the discretization time-step used in most simulations has been 5 ms. This decision has been conditioned by the fact that the absolute refractory period in Aplysia is 10 ms, which makes it impossible for the neuron to fire at frequencies higher than 100 Hz.

Figure 3.2 shows the autonomous behavior of the model for m_{Pb_1}=-40 mv and σ_{Pb_1}=1.5 mv, as well as its response to different stimulation patterns.

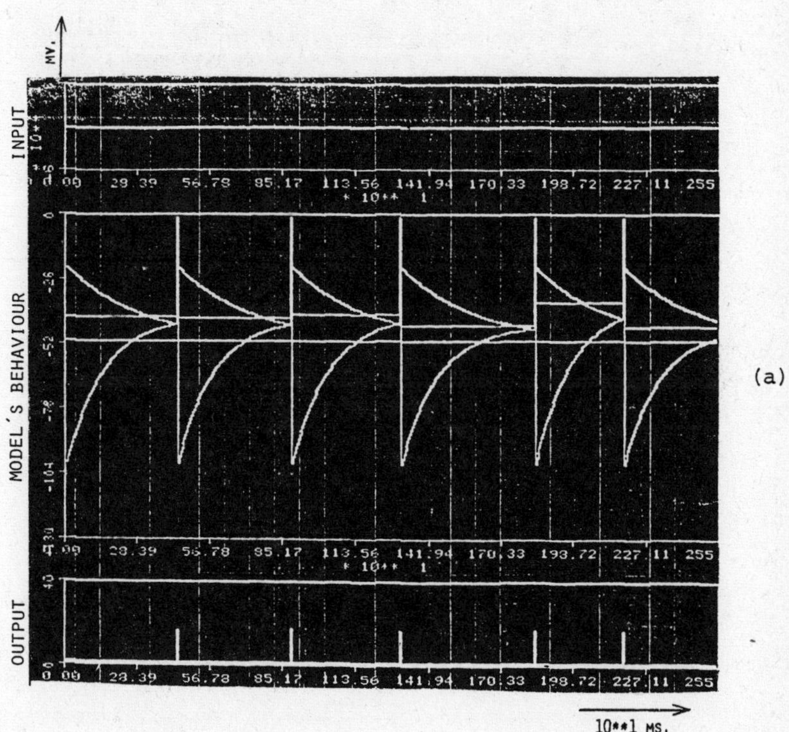

(a)

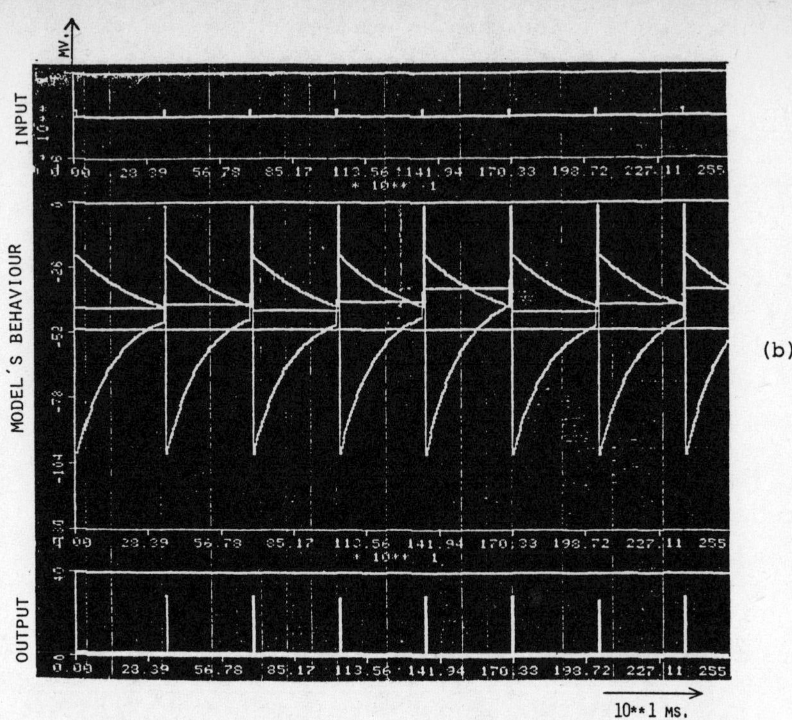
MV.
INPUT
MODEL'S BEHAVIOUR
OUTPUT
10**1 MS.

(b)

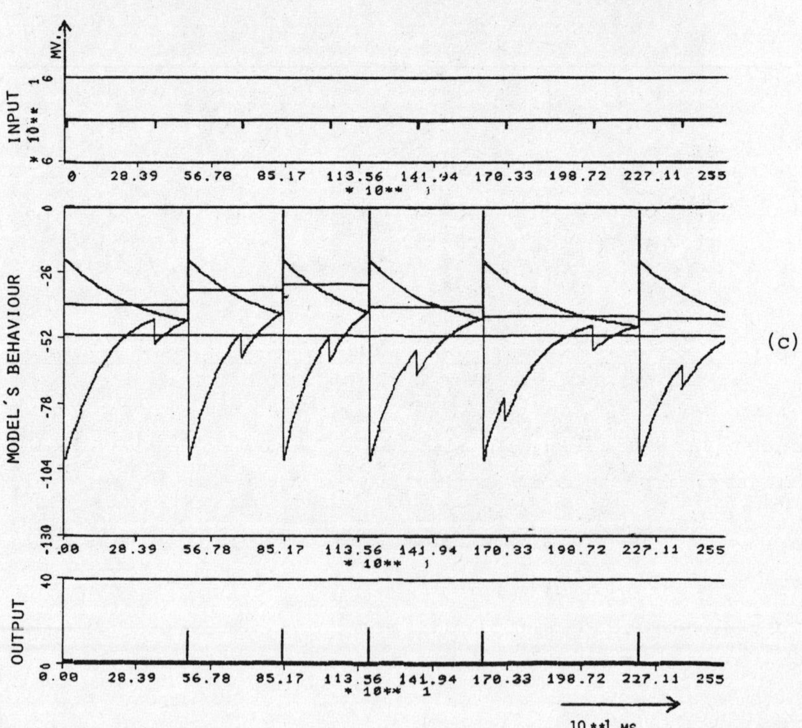
MV.
INPUT
MODEL'S BEHAVIOUR
OUTPUT
10**1 MS.

(c)

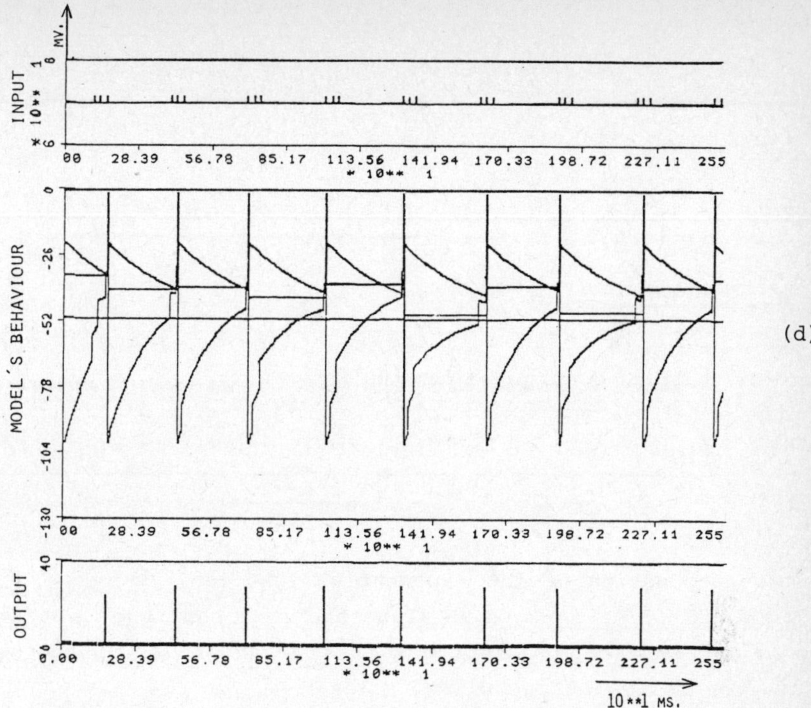

*** Figure 3.2 - Behavior of the model in four different stimulation conditions: (a) in the absence of stimulation; (b) submitted to a pulsed excitatory stimulus of period 350 ms; (c) submitted to a pulsed inhibitory stimulus of identical period; and (d) submitted to trains of excitatory impulses every 300 ms. Observe in (a) the interspike interval variability determined by changes in the asymptotic limit of the spontaneous membrane potential. In (b) and (d), synchronized and quasi-synchronized entrainment are attained, respectively, the output amplitude being higher than in the other two situations. In (c), non-synchronized entrainment is only partly attained.

3.3.1 - Variability of the spontaneous interspike interval

Since the length of the interspike interval T_s is a function of the random variable Pb_1, T_s itself becomes a random variable and, as we will see, its probability density function can be deduced from that of Pb_1.

The following exponential polynomic equation, resulting from equating the value of the spontaneous membrane potential Pb with that of the threshold H at the time of firing T_s, expresses implicitly the dependence of T_s on Pb_1:

$$H_1+(H_0-H_1) \cdot e^{-\tau_h \cdot T_s} = Pb_1+(Pb_0-Pb_1) \cdot e^{-\tau_b \cdot T_s} \tag{3.8}$$

To calculate the probability density function of T_s, Pb_1 has to be expressed as an explicit function of T_s:

$$Pb_1 = h(T_s) = \frac{H_1+(H_0-H_1) \cdot e^{-\tau_h \cdot T_s} - Pb_0 \cdot e^{-\tau_b \cdot T_s}}{1-e^{-\tau_b \cdot T_s}} \tag{3.9}$$

The function h is defined in the interval $(0, \infty)$ and is continuous, derivable and has continuous derivative:

$$\frac{dh}{dT_s} = \frac{\tau_h (H_1-H_0) e^{-\tau_h \cdot T_s}(1-e^{-\tau_b \cdot T_s}) + \tau_b e^{-\tau_b \cdot T_s}(Pb_0-H_1-(H_0-H_1) e^{-\tau_h T_s})}{(1-e^{-\tau_b T_s})^2} \tag{3.10}$$

From the properties of the exponential function, certain restrictions of the model ($\tau_h>0$, $\tau_b>0$, $H_0>H_1$) and the fact that the membrane potential is an increasing function with time and thus, up to the time of firing, the following inequality holds:

$$Pb_0 < H_1 + (H_0-H_1) \; e^{-\tau_h \cdot T_s}$$

we deduce that:

$$\frac{dh}{dT_s} < 0$$

Therefore, the function h is strictly decreasing and, consequently, one-to-one. Its image is (H_1, ∞).

Particularizing equations 3.9 and 3.10 to the case of the model simulated, we obtain:

$$Pb_1 = h(T_s) = \frac{-50+30 \; e^{-3.43 \; T_s} +100 \; e^{-6.93 \; T_s}}{1-e^{-6.93 \; T_s}} \tag{3:11}$$

$$\frac{dh}{dT_s} = \frac{-102.9 \; e^{-3.43 \; T_s} -346.5 \; e^{-6.93 \; T_s} -36.4 \; e^{-10.36 \; T_s}}{(1-e^{-6.93 \; T_s})^2} \tag{3.12}$$

Figure 3.3 shows a graphical representation of these two functions.

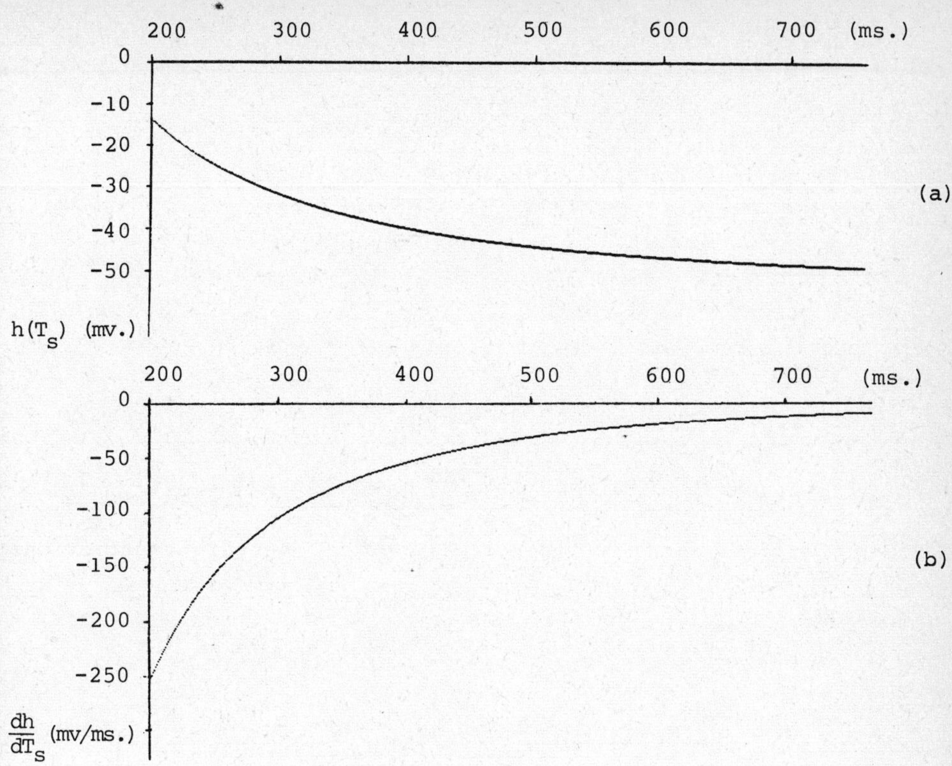

*** Figure 3.3 - Graphical representation of the functions: (a) $h(T_s)$ and (b) dh/dT_s.

Now, we will compute the probability density function of T_s. Its distribution function is given by:

$$F_{T_s}(a) = \text{Prob}\left[T_s \varepsilon (-\infty, a]\right] = \text{Prob}\left[Pb_1 \varepsilon\ h(-\infty, a]\right]$$

and since h is strictly decreasing:

$$F_{T_s}(a) = \text{Prob}\left[Pb_1 \varepsilon \left[h(a), \infty\right)\right] = 1 - F_{Pb_1}(h(a))$$

where F_{Pb_1} is the probability distribution of Pb_1, which being Gaussian is derivable:

$$F_{T_s}(a) = \int_{h(a)}^{\infty} f_{Pb_1}(x)\, dx$$

or

$$f_{T_s}(a) = -f_{Pb_1}(h(a)) \frac{dh}{dT_s}(a) \qquad (3.13)$$

Substituting f_{Pb_1} for its Gaussian expression and expliciting the functions h and dh/dT_s according to equations 3.9 and 3.10, we obtain:

$$f_{T_s}(a) = \frac{-1}{\sqrt{2\pi}\,\sigma_{Pb_1}} e^{-\left[\frac{\left(\frac{H_1+(H_0-H_1)e^{-\tau_h \cdot a}-Pb_0 e^{-\tau_b \cdot a}}{1-e^{-\tau_b \cdot a}} - m_{Pb_1}\right)^2}{2\,\sigma^2_{Pb_1}}\right]} \cdot$$

$$\cdot \frac{H_1\tau_h\, e^{-\tau_h a} + (Pb_0-H_1)\tau_b\, e^{-\tau_b a} - (H_1\tau_h + (H_0-H_1)\tau_b)e^{-(\tau_h+\tau_b)a}}{(1-e^{-\tau_b \cdot a})^2} \qquad (3.14)$$

Setting the parameters, except σ_{Pb_1}, to the values used in the simulation:

$$f_{T_s}(a) = \frac{-1}{\sqrt{2\pi}\,\sigma_{Pb_1}} e^{-\left[\frac{\left(\frac{-50+30\,e^{-3.43\,a}+100\,e^{-6.93\,a}}{1-e^{-6.93\,a}} + 40\right)^2}{2\,\sigma^2_{Pb_1}}\right]} \cdot$$

$$\cdot \frac{-102.9\,e^{-3.43\,a} - 346.5\,e^{-6.93\,a} - 36.4\,e^{-10.36\,a}}{(1-e^{-6.93\,a})^2} \qquad (3.15)$$

Figure 3.4 allows comparison of the histogram of spontaneous interspike intervals, obtained through simulation, with the probability density function f_{T_s}, obtained analytically, for different values of σ_{Pb_1}. The curve that results is qualitatively similar to the hyperbolic normal, pointed out by some experimental studies as the one that is in good agreement with the data obtained (Junge and Moore, 1966; Tuckwell, 1978).

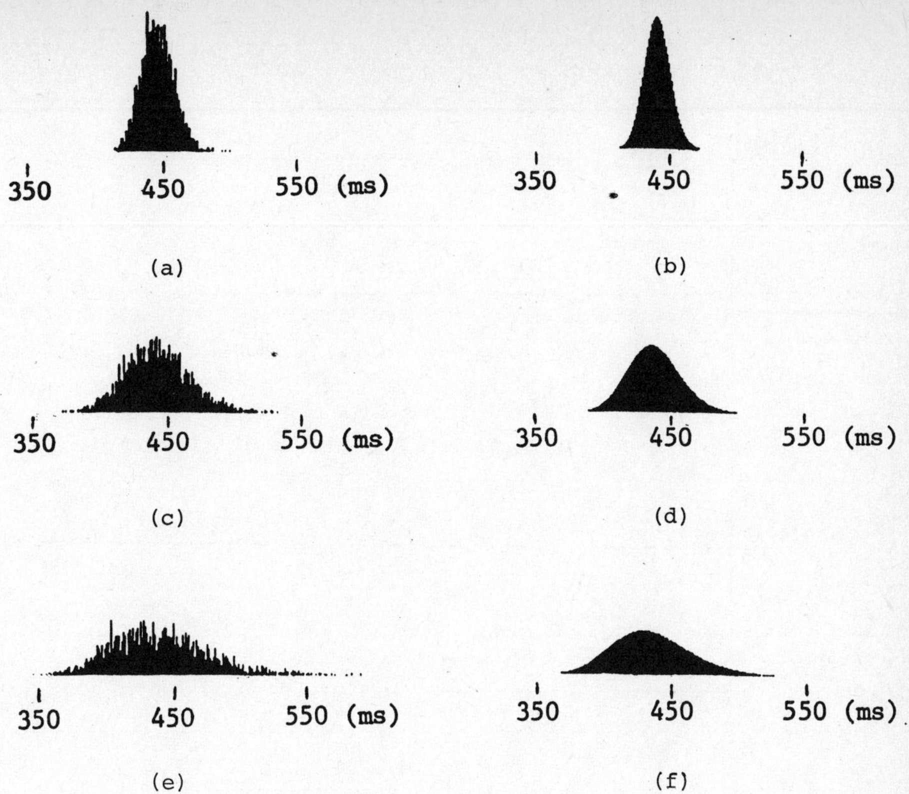

*** Figure 3.4 - Comparison between the histogram of spontaneous interspike intervals obtained through simulation (left column) and the probability density function obtained analytically (right column), for the following values of σ_{Pb_1}: (a) and (b) σ_{Pb_1}=0.5 mv; (c) and (d) σ_{Pb_1} =1.0 mv; (e) and (f) σ_{Pb_1}=1.5 mv.

In relation to the quantitative data, there seems to be a certain consensus in the measurement of a coefficient of variation of the order of 0.05 for an average firing frequency of 2.5 Hz. Since f_{T_S} is a function of σ_{Pb_1}, the mathematical expectation of T_S and its standard deviation depend on σ_{Pb_1}. In view of the difficulty that the integral computation of these indicators presents, we have opted for its discrete approximation (Figure 3.5). It follows from the graphics that the assignment σ_{Pb_1}=1.0 mv is a good choice for the purpose of concordance with the experimental data.

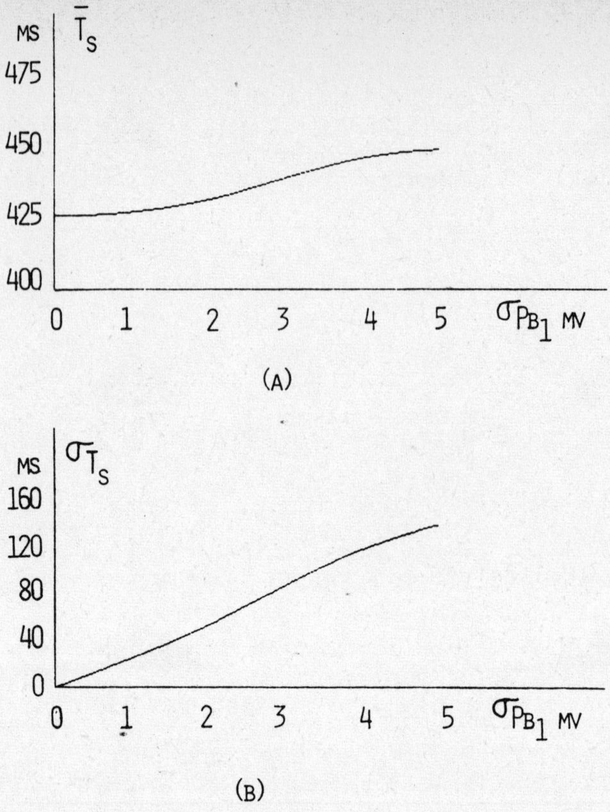

*** Figure 3.5 - (a) Mathematical expectation and (b) standard deviation of the average interspike interval as a function of σ_{Pb_1}. Dividing σ_{T_S} by $\bar{T}_S$ we obtain the coefficient of variation of T_S, which is about 0.05 for σ_{Pb_1}=1 mv.

3.3.2 - Response to occasional perturbations

In Section 2.1.1.2 the Phase Response Curve (PRC) has been defined and its dependence on the amplitude of the stimulation, in the case of pacemaker neurons, has been indicated. To derive the analytic expression of $\delta(\varnothing,m)$ for the proposed model, we suppose that T_S is the spontaneous interspike interval and T'_S is the interval modified by the arrival of an input impulse of amplitude m at the phase $\varnothing$, then:

$$\delta(\varnothing, m) = \frac{T'_S}{T_S} - 1 \qquad (3.16)$$

On the other hand, equating the value of the threshold with that of the membrane potential at the time of firing, we have:

$$H_1+(H_0-H_1).e^{-\tau_h \cdot T_s'} = Pb_1+(Pb_0-Pb_1)\ e^{-\tau_b \cdot T_s'}+ m.e^{-\tau_s (T_s'-\emptyset . T_s)} \quad (3.17)$$

and substituting T_s' for its value obtained in equation 3.16:

$$H_1+(H_0-H_1).e^{-\tau_h \cdot T_s (1+\delta(\emptyset,m))} =$$

$$= Pb_1+(Pb_0-Pb_1).e^{-\tau_b \cdot T_s (1+\delta(\emptyset,m))} + m.e^{-\tau_s \cdot T_s (1+\delta(\emptyset,m)-\emptyset)} \quad (3.18)$$

For each point $(\emptyset,m)$, δ is a random variable, since it depends on Pb_1.

Experimental studies usually report only the average value of this variable at each point or, even less, the result of carrying out a regression on these data (Section 2.1.1.2). To make it possible to appreciate the proper behavior of the model in relation to this characterization, we have included in Figure 3.6(a) the PRCs --in terms of the mean value of $\delta(\emptyset,m)$-- corresponding to six stimulation amplitudes, three excitatory and three inhibitory ones. The values of the parameters used in the simulation have been the ones determined in the preceding section: m_{Pb_1}=-40 mv and σ_{Pb_1}=1 mv.

For comparative purposes, and anticipating their use in the next section, the PRCs corresponding to the same stimulation amplitudes as above, for the deterministic version of the model (σ_{Pb_1}=0 mv), are shown in Figure 3.6(b).

Note that the excitatory stimuli always produce an advancement of phase, while the inhibitory ones provoke a delay. During the absolute refractory period, the phase remains unaltered. The segment of slope approximately +1, that appears at the end of the cycle in the PRCs corresponding to excitatory stimuli, arises because the neuron is forced by the stimulus to fire without delay. In Winfree's classification, described in Section 2.1.1.2, these PRCs correspond to an oscillator of Type 1, while the ones generated by inhibitory stimuli are unclassifiable, since they do not give rise to a continuous curve on the torus.

If the absolute refractory period were suppressed from the model, it would behave as an oscillator of Type 1, for excitatory stimuli of amplitude less than (H_0-Pb_0) mv, and as an oscillator of Type 0, for stimuli of amplitude greater than this value. The singularity commented on in Section 2.1.1.2 would arise then at the point $(\varnothing,m)=(0,H_0-Pb_0)$, as can be seen in Figure 3.7.

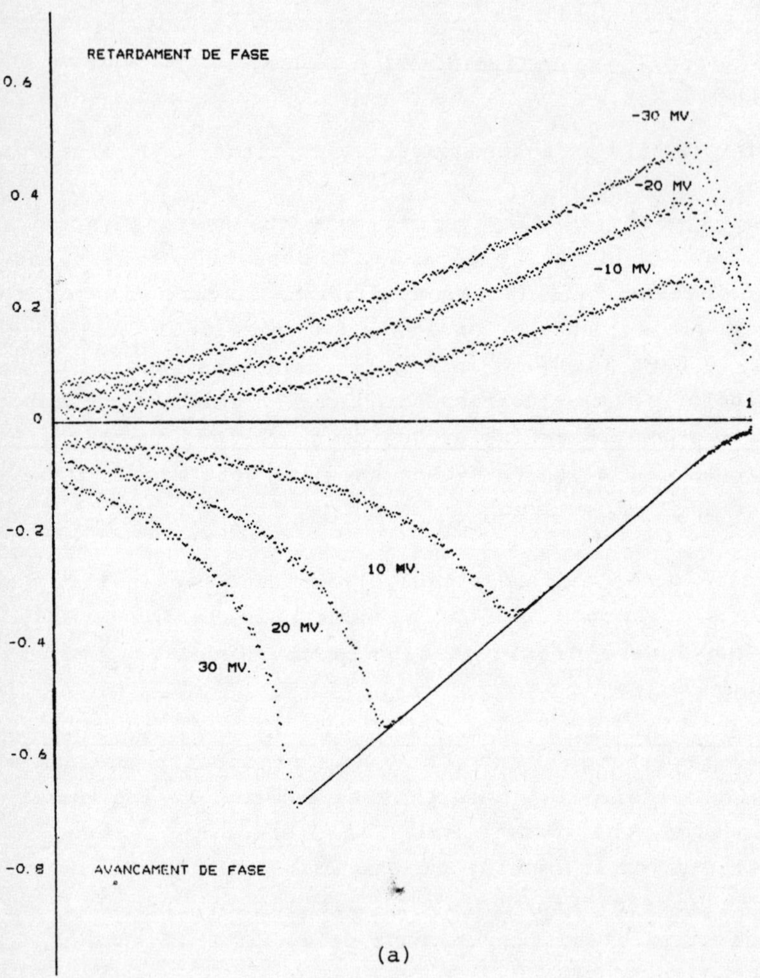

(a)

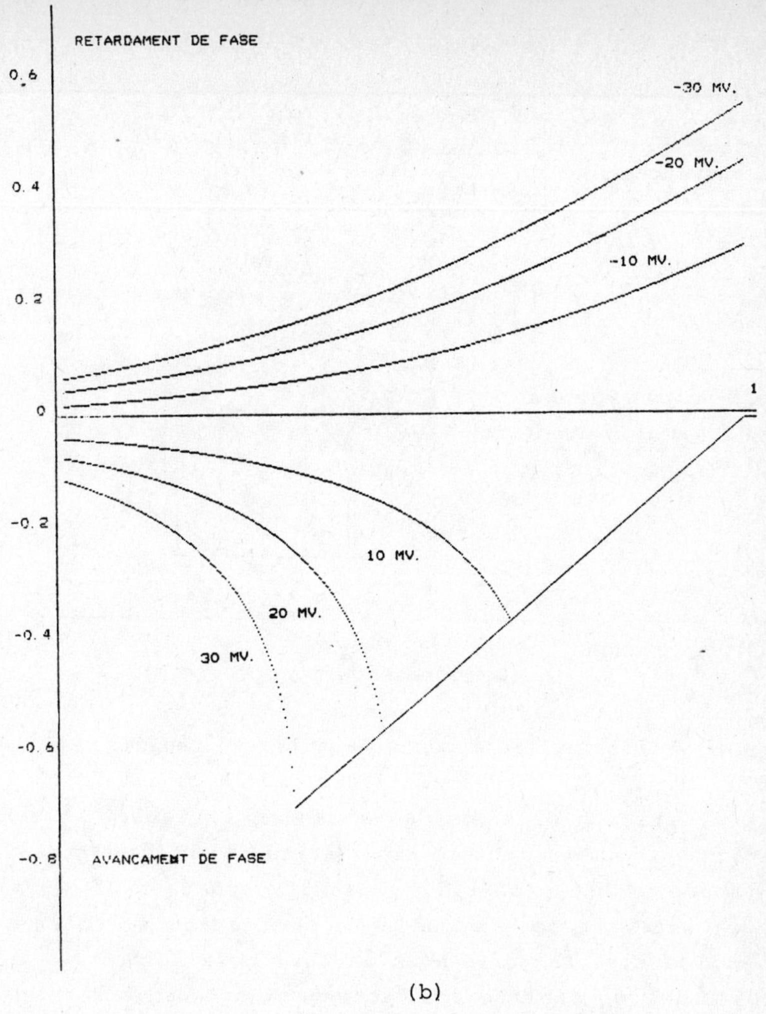

(b)

*** Figure 3.6 - PRCs for three amplitudes of excitatory stimulation (10 mv, 20 mv and 30 mv) and three amplitudes of inhibitory stimulation (-10 mv, -20 mv and -30 mv): (a) σ_{Pb_1}=1 mv; (b) σ_{Pb_1}=0 mv. The scatter appearing in (a), but not in (b), for phases close to 1 is an artifact due to the interspike interval variability. As in experimental studies (compare especially with Figure 2.4(d)), the phases at which input impulses have to be delivered are computed taking as reference the average interspike interval. For short interspike interval instances, input impulses arrive after discharge, provoking neither the predicted phase-delay in the inhibitory cases, nor the predicted phase-advancement in the excitatory ones. See text for more details.

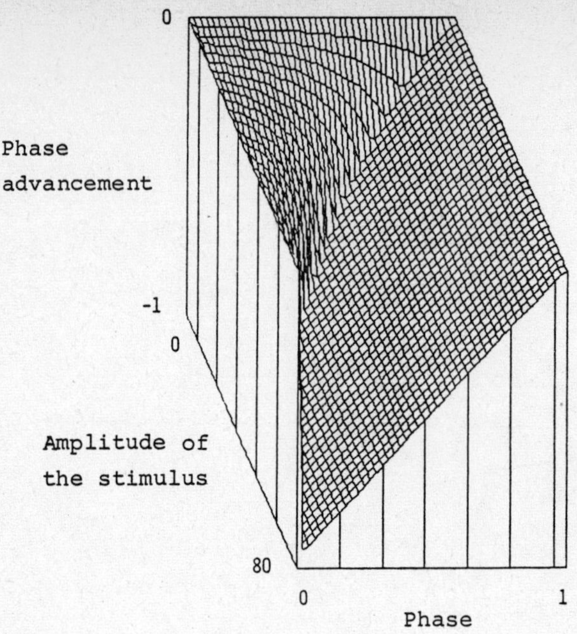

*** Figure 3.7 - 3-D representation of $\delta(\varnothing,m)$.

A new way of validating the aspects relative to randomness, not only of this model, but of any nondeterministic neuron oscillator, is provided by the consideration of $\delta(\varnothing,m)$ as a random variable. It consists in comparing, not the mathematical expectation, as we have done above, but the very shape of the probability density function of $\delta(\varnothing,m)$ in each point $(\varnothing,m)$ with the corresponding experimental histogram, through several indicators. There is no sign that this method has been used, probably because of the huge collecting of intensive data (histograms for each phase and amplitude) and extensive data (in all the space of phases and amplitudes) required. The simulations carried out to explore the fluctuations of the standard deviation of the probability density of $\delta(\varnothing,m)$ when varying $\varnothing$ and m (Figure 3.8) point toward the prediction, to be confirmed or refuted experimentally, that the mentioned deviation is an increasing function with respect to phases less than λ (phase at which the maximum phase-advancement takes place), and practically nil for phases greater than this value, except for those just below 1. An increase in m leads to a decrease in λ and, in general, contributes to make more extreme the dispersion in the zones most prone to it --especially those below λ-- and to diminish it in the remaining ones --especially those below 1.

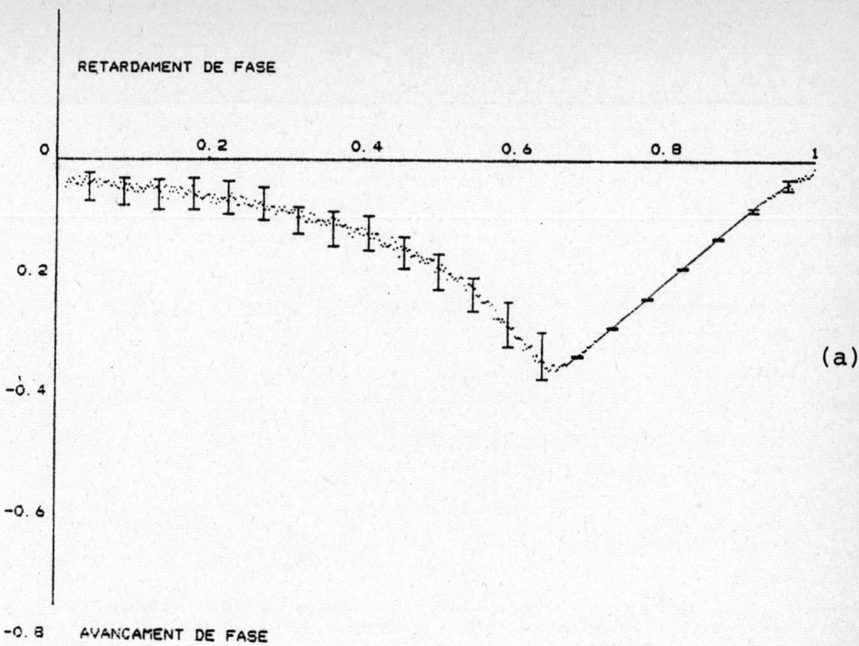
RETARDAMENT DE FASE
0
0.2
0.4
0.6
0.8
1
0.2
-0.4
-0.6
-0.8
AVANCAMENT DE FASE
(a)

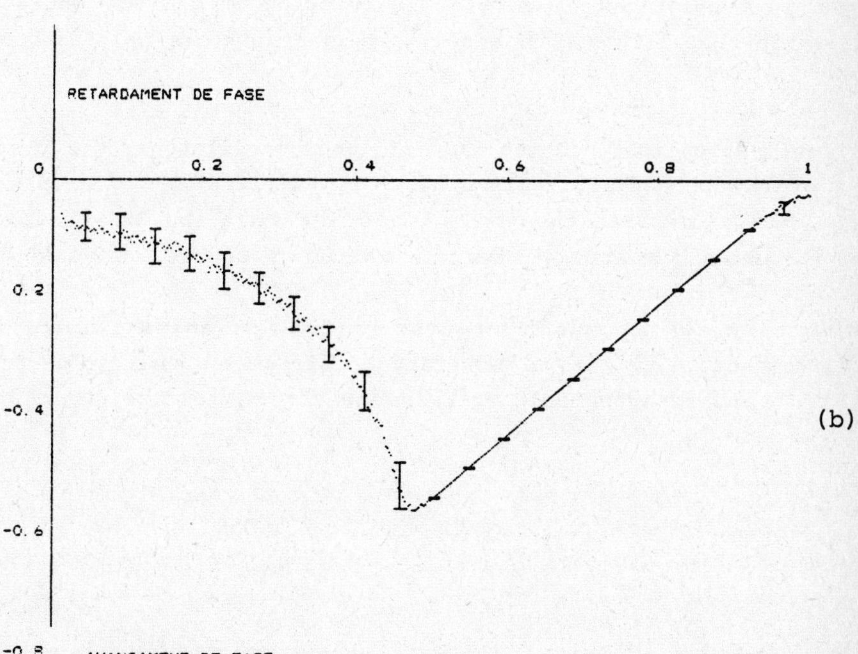
RETARDAMENT DE FASE
0
0.2
0.4
0.6
0.8
1
0.2
-0.4
-0.6
-0.8
AVANCAMENT DE FASE
(b)

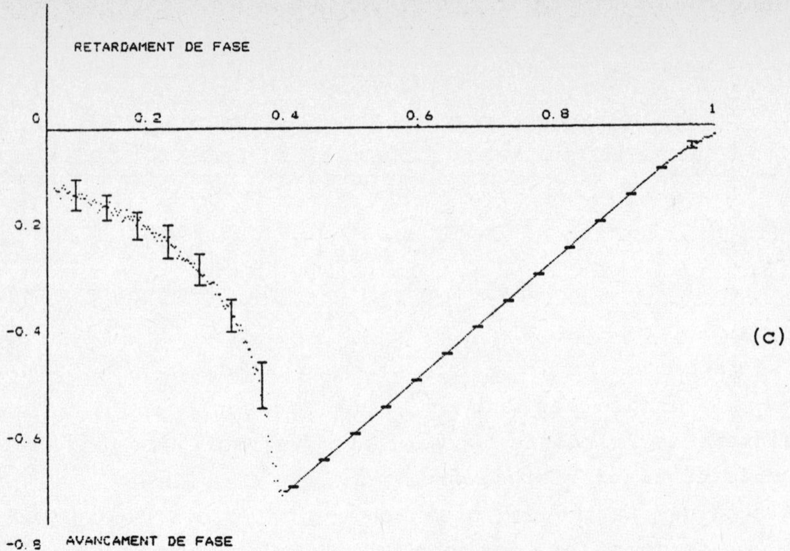

*** Figure 3.8 - Fluctuation of the standard deviation (I) of $\delta(\emptyset,m)$, when varying $\emptyset$ and m. (a) m=10 mv. (b) m=20 mv. (c) m=30 mv.

3.3.3 - Entrainment

Since the next chapter is dedicated completely to entrainment, this section will only contain a brief description of the most outstanding characteristics of this phenomenon, observed in the simulations carried out.

We first mention.that, in the case of ideal oscillators, the **Theory of Oscillators**, paradigmatically applied by Pavlidis (1973) to the biological field, provides the tools to derive from the PRC the bands of input frequencies that yield certain ratios of stable entrainment.

Specifically, if T_0 and T_1 are the periods of an oscillator and an input, respectively, then a necessary condition for one-to-one entrainment (k=1) and superharmonic entrainment ($1<k\in N$) to take place is that:

$$T_0(k+\min_{\emptyset\varepsilon I}\delta(\emptyset,m) \leqslant T_1 \leqslant T_0(k+\max_{\emptyset\varepsilon I}\delta(\emptyset,m)), \tag{3.19}$$

$$\forall I=[a,b]=\{\emptyset\varepsilon[0,1] : 0 < \frac{d\delta(\emptyset,m)}{d\emptyset} < 2\},\ \forall m\varepsilon\ \mathbb{R}$$

with the understanding that the bands of stimulation frequencies thus defined depend on the amplitude of stimulation and that, for each amplitude, frequencies that correspond to different bands can give place to the same entrainment ratio. The cyclic pattern of interspike intervals imposed on the oscillator will be $T_0(1-\delta(\emptyset,m))$ in (1:1) entrainment, $(T_0,T_0(1-\delta(\emptyset,m)))$ in (1:2) entrainment, $(T_0,T_0,T_0(1-\delta(\emptyset,m)))$ in (1:3) entrainment, etc.

Applying equation 3.19 to the PRCs of Figure 3.6, Tables 3.2 and 3.3 are obtained. In this case, a single band of frequencies corresponds to each entrainment ratio for each stimulation amplitude, since each curve contains a single segment of slope between 0 and 2. Attainment of entrainment is immediate for excitatory stimuli, as long as they arrive in the interval of slope approximately +1; while for inhibitory ones, it is attained at the end of a monotonically convergent process, since the slope is comprised between 0 and 1. For completeness, we mention that convergence is oscillatory for slopes between 1 and 2. In the excitatory case, a stimulus-neuron synchronization arises; in the inhibitory one, entrainment is non-synchronized.

ENTRAINMENT BANDS

AMPLITUDE OF THE STIMULUS	one-to-one	one-to-two	one-to-k
30 mv	$151 \leqslant T_1 \leqslant 430$	$581 \leqslant T_1 \leqslant 860$	$430k - 279 \leqslant T_1 \leqslant 430k$
20 mv	$209 \leqslant T_1 \leqslant 430$	$639 \leqslant T_1 \leqslant 860$	$430k - 221 \leqslant T_1 \leqslant 430k$
10 mv	$289 \leqslant T_1 \leqslant 430$	$719 \leqslant T_1 \leqslant 860$	$430k - 141 \leqslant T_1 \leqslant 430k$
-10 mv	$439 \leqslant T_1 \leqslant 535$	$869 \leqslant T_1 \leqslant 965$	$430k + 9 \leqslant T_1 \leqslant 430k + 105$
-20 mv	$453 \leqslant T_1 \leqslant 590$	$883 \leqslant T_1 \leqslant 1020$	$430k + 23 \leqslant T_1 \leqslant 430k + 160$
-30 mv	$456 \leqslant T_1 \leqslant 630$	$886 \leqslant T_1 \leqslant 1060$	$430k + 26 \leqslant T_1 \leqslant 430k + 200$

*** Table 3.2 - Bands of one-to-one and superharmonic entrainment, as a function of the stimulation amplitude, for the model with $m_{Pb_1}=-40$ mv and $\sigma_{Pb_1}=1$ mv.

ENTRAINMENT BANDS

AMPLITUDE OF THE STIMULUS	one-to-one	one-to-two	one-to-k
30 mv	$146 \leqslant T_1 \leqslant 430$	$576 \leqslant T_1 \leqslant 860$	$430k + 284 \leqslant T_1 \leqslant 430k$
20 mv	$200 \leqslant T_1 \leqslant 430$	$630 \leqslant T_1 \leqslant 860$	$430k - 230 \leqslant T_1 \leqslant 430k$
10 mv	$280 \leqslant T_1 \leqslant 430$	$710 \leqslant T_1 \leqslant 860$	$430k - 150 \leqslant T_1 \leqslant 430k$
-10 mv	$437 \leqslant T_1 \leqslant 551$	$867 \leqslant T_1 \leqslant 981$	$430k + 7 \leqslant T_1 \leqslant 430k + 121$
-20 mv	$447 \leqslant T_1 \leqslant 612$	$877 \leqslant T_1 \leqslant 1042$	$430k + 17 \leqslant T_1 \leqslant 430k + 182$
-30 mv	$457 \leqslant T_1 \leqslant 653$	$887 \leqslant T_1 \leqslant 1083$	$430k + 27 \leqslant T_1 \leqslant 430k + 223$

*** Table 3.3 - Bands of one-to-one and superharmonic entrainment, as a function of the stimulation amplitude for the deterministic version of the model (σ_{Pb_1} =0 mv).

Empirically, either in experimentation or simulation, the bands and types of entrainment can be studied by stimulating the neuron with a wide band of frequencies and representing the histogram of interspike intervals as a function of the period of the stimulus (Pinsker and Ayers, 1983). Figures 3.9 to 3.12 show a set of graphics of this kind; the way they are to be read is explained in the caption of Figure 3.9.

Obviously, the simulation results for the deterministic version of the model (Figure 3.9) fully match the analytic predictions relative to the entrainment bands and types of convergence. The entrainment bands, both in the excitatory and the inhibitory side, widen progressively as the stimulation amplitude increases. The histograms that characterize (1:2)* entrainment show two peaks of equal height: one that corresponds to the spontaneous period of the neuron and the other, to the difference between the period of the stimulus and that of the neuron; the cyclic pattern generated is:

$$(T_0, T_0(1-\delta(\emptyset,\ m))) = (T_0, T_0.\emptyset) = (T_0, T_1 - T_0)$$

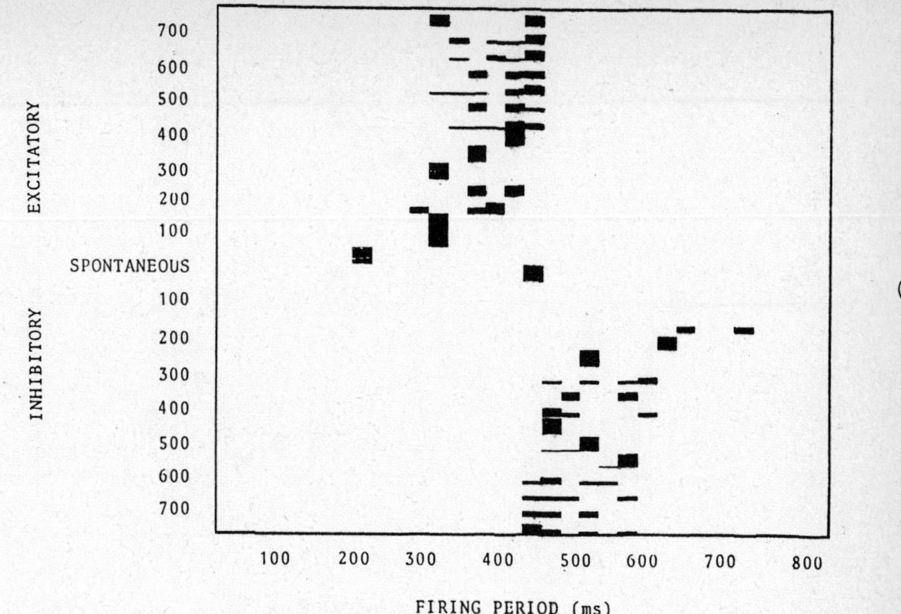
STIMULATION PERIOD (ms)
EXCITATORY
INHIBITORY
700
600
500
400
300
200
100
SPONTANEOUS
100
200
300
400
500
600
700
100
200
300
400
500
600
700
800
FIRING PERIOD (ms)

(a)

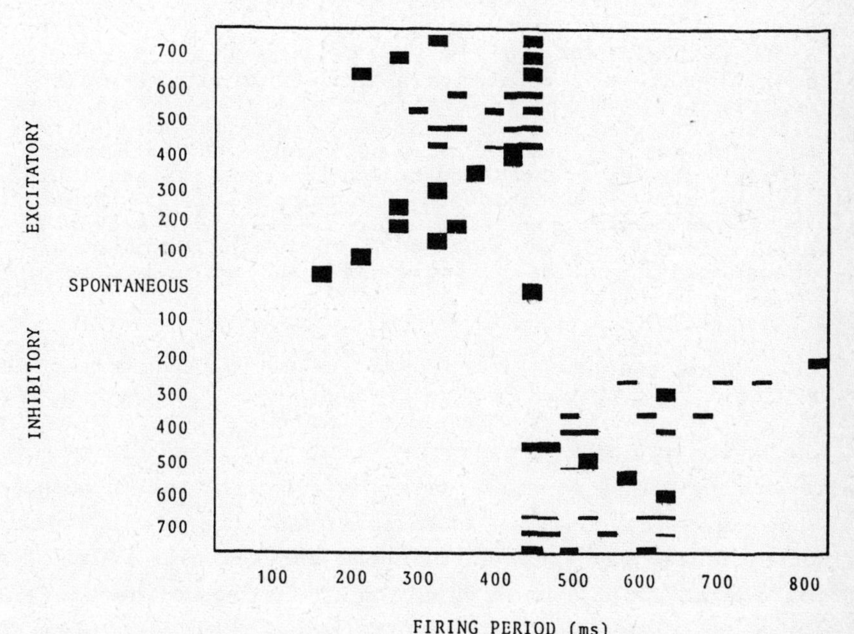
STIMULATION PERIOD (ms)
EXCITATORY
INHIBITORY
700
600
500
400
300
200
100
SPONTANEOUS
100
200
300
400
500
600
700
100
200
300
400
500
600
700
800
FIRING PERIOD (ms)

(b)

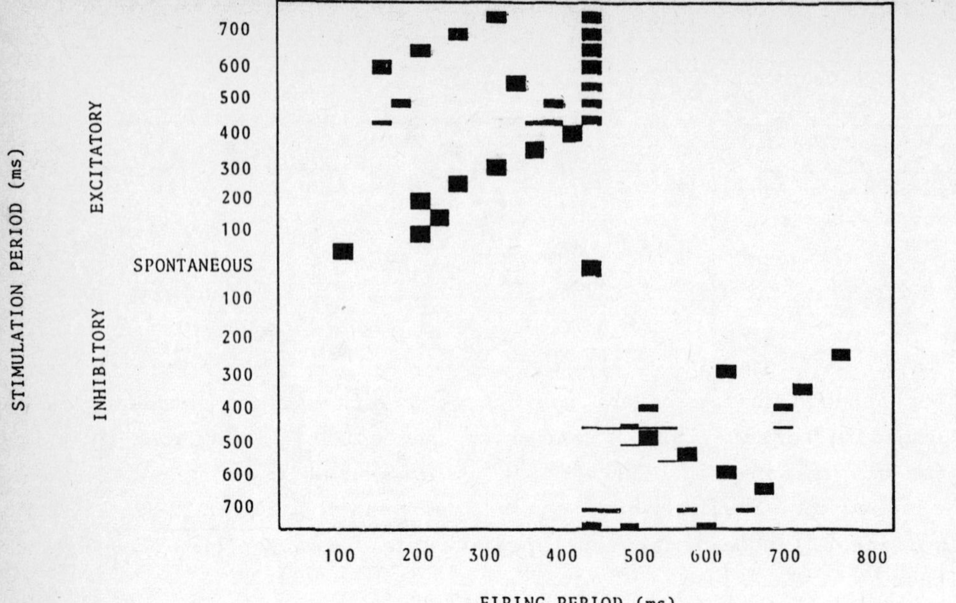

*** Figure 3.9 - Histograms of interspike intervals as a function of the period of the excitatory stimulation (up) and the inhibitory stimulation (down), for $\sigma_{Pb_1}=0$ mv and with the learning rule turned off. (a) m=10 mv, (b) m=20 mv, (c) m=30 mv. In the axis of abscissas, the neuron's interspike intervals between 25 and 800 ms are represented, with a discretization step of 25 ms. The origin of the axis of ordinates corresponds to spontaneous behavior; above it, the periods of the excitatory stimuli are represented and below it, those of the inhibitory stimuli, both covering, in opposite directions, the spectrum that goes from 50 to 750 ms, with a discretization step of 50 ms. The entrainment regions are characterized by the lining up of a series of maxima of consecutive histograms.

Figure 3.10 shows the same kind of data resulting from the simulation of the proposed model with $m_{Pb_1}=-40$ mv and $\sigma_{Pb_1}=1$ mv. On the excitatory side, the bands of one-to-one entrainment predicted are maintained, with only a slight loss of stimulus-neuron synchronization being observed in the extremes. On the inhibitory side, on the contrary, the coordination is largely lost, entrainment showing only in the tendency of histograms to adopt a configuration in stripes of negative slope, qualified as "paradoxical" in the literature. This qualification is because, in principle, the period of the neuron should shorten with the decrease of the quantity of inhibition furnished; that is to say, with the widening of the stimulation period and, as a result, the slope should be positive. The uneven incidence of randomness in the excitatory and inhibitory cases derives from the different way entrainment is attained in each case: immediately, in the excitatory case, and following a monotonically convergent process, in the inhibitory one. In both, nevertheless, an increase in the amplitude of the stimulation improves the quality of entrainment and widens the region in which it occurs.

Other aspects that a priori could have an effect upon entrainment were studied in Torras (1981), leading to the conclusion that the initial phase of presentation of the stimulus affects neither the existence nor the speed of attainment of entrainment, and that the latter is not significantly modified by the incorporation of a source of randomness, either.

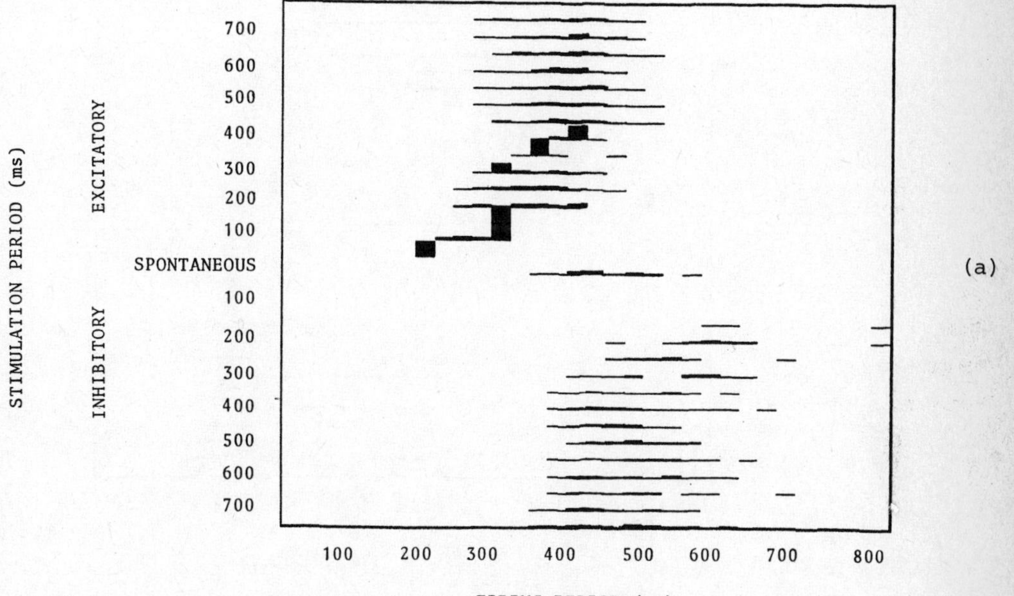

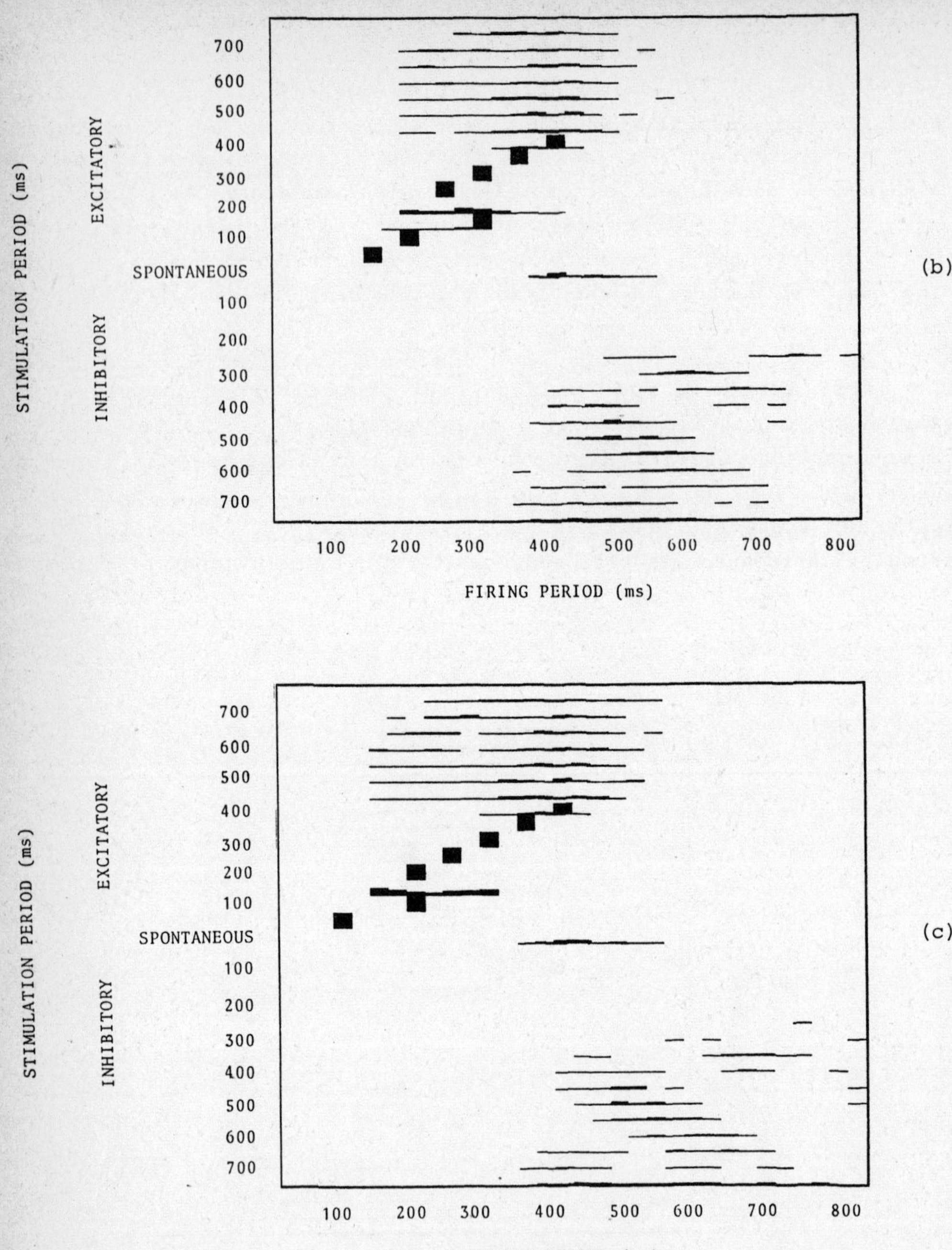

*** Figure 3.10 - Histograms of interspike intervals as a function of the stimulation period, for σ_{Pb_1}=1 mv and with the learning rule turned off. (a) m=10 mv. (b) m=20 mv. (c) m=30 mv.

3.3.4 - Learning

"A cat that once sat on a hot stove will never again sit on a hot stove. Or on a cold one either".

M. Twain.

The results and conclusions derived from the simulation of the model, equipped successively with each of the three learning rules defined in Section 3.2.1, are outlined below.

The first learning rule only behaves properly when the neuron is excited at frequencies slightly higher than its own or than some of its harmonics. More specifically, the neuron learns the stimulation frequencies that yield one-to-one entrainment and do not differ from the spontaneous one by more than 30%, as well as those harmonics of the stimulation frequencies yielding superharmonic entrainment that do not differ from the spontaneous one by more than a percentage which diminishes as the harmonic increases.

In the remaining circumstances, the behavior of the model diverges greatly from that observed in the experimental preparations described in Section 2.1.1.4. The excitatory stimuli always shorten the interspike interval and the inhibitory ones, lengthen it. Stationarity can only be attained in two ways: because m_{Pb_1} has reached its upper or lower bound, or because synchronization has been attained. Apart from the mentioned favorable cases, stationarity is only possible for frequencies close to the spontaneous one or its harmonics that do not yield entrainment. Whether it is attained depends on the initial phase of presentation of the stimulus.

The incorrect behavior of this rule follows from an improper use of the goal-driven strategy. It makes no sense to fix objectives that are valid under certain conditions and pursue them under different conditions, as it makes no sense, either, to seek objectives whose approximation leads necessarily to distancing from them. Both maxims are applicable to this case: the objectives are fixed for a situation free of stimulation and are sought under stimulation, which determines their variation after each discharge, except when firing occurs synchronously with the stimulus.

What has been said, along with the low biological plausibility of this rule, commented on in Section 3.2.2, is reason enough to discard further use of this rule in the network.

Figure 3.11 shows the results of simulating the model equipped with **the second learning rule.** Comparing the two graphics of this figure with those in Figures 3.9(a) and 3.10(a), we see that the neuron learns to synchronize with excitatory frequencies within a wide band around the spontaneous one. Actually, it not only learns to synchronize, but, when the stimulation disappears, the neuron continues firing at the frequency of the stimulus. Table 3.4 shows the learned frequencies, in terms of the average spontaneous interspike interval, as a function of the period of the excitatory stimulus. The results confirm a certain learning of the harmonics, that was already beginning to appear in Figure 3.11(b), shortening the region where it occurs with the increase of the harmonic. The above results correspond to the model with the parameter P^{*}=4 mv. The range of learned frequencies tends toward the lower ones or the higher ones, depending on whether P^{*} increases or decreases. Inhibitory stimulation, as has been commented on when defining the rule, has no effect upon learning.

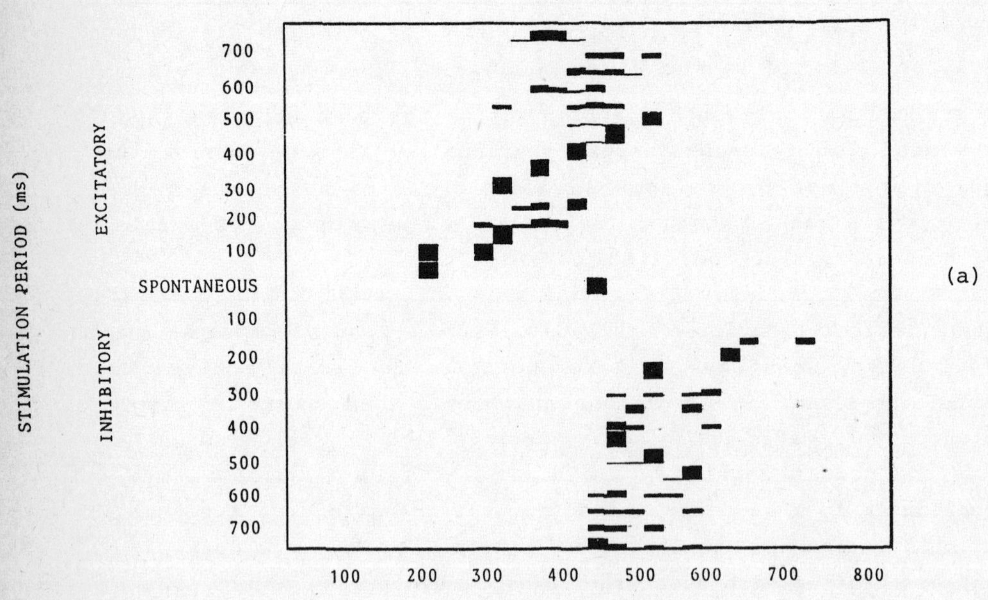

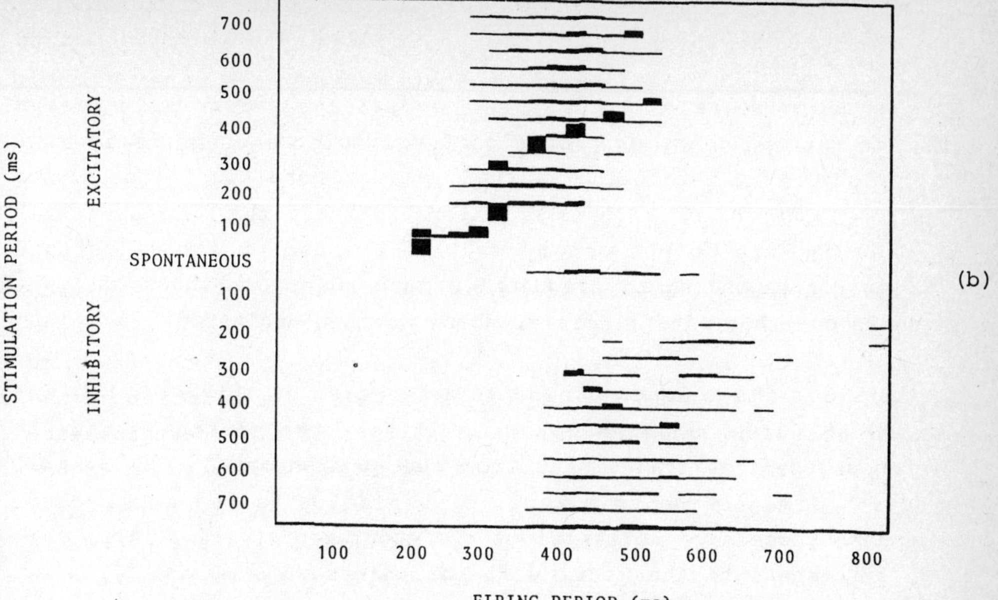

*** Figure 3.11 - Histograms of interspike intervals as a function of the period of a stimulus of amplitude 10 mv, for the model equipped with the second learning rule. (a) σ_{Pb_1}=0 mv. (b) σ_{Pb_1}=1 mv.

Period of the stimulus (ms)	Learned period (ms)	Period of the stimulus (ms)	Learned period (ms)	Period of the stimulus (ms)	Learned period (ms)
50	---	550	---	1050	---
100	---	600	---	1100	---
150	---	650	---	1150	---
200	---	700	---	*	---
250	---	*	---	1200	400
*	---	750	375	*	408
300	300	*	387	1250	415
*	325	800	400	*	425
350	350	*	412	1300	435
*	375	850	425	*	---
400	400	*	438	1350	---
*	425	900	450	1400	---
450	450	*	---	1450	---
*	475	950	---	1500	---
500	500	1000	---		
*	---				

*** Table 3.4 - Period learned by the neuron model equipped with the second learning rule, as a function of the period of the stimulus presented.

Even though this second rule yieds a remarkable improvement over the first one, since it permits learning excitatory frequencies lower than the spontaneous one and attains stationarity without requiring m_{Pb_1} to adopt its maximum or minimum value, its use for the learning of external stimuli in the network has also been rejected, because of its lack of concordance with the experimental data in a crucial aspect: the excessively wide range of frequencies of which it permits learning.

The third learning rule (Figure 3.12) excels the second rule in two senses: on the one hand, the range of frequencies around the spontaneous one susceptible to being learned is much more highly restricted and, on the other hand, certain collateral effects that appeared with the preceding rule do not occur, such as the distortion of certain histograms corresponding to frequencies much lower or higher than the spontaneous one that, in principle, should remain unaltered.

Table 3.5 shows the learned period as a function of the period of the presented stimulus. One deduces from it that the neuron is able to learn imposed periods that differ from the spontaneous one by no more than 50 ms, diminishing this range of variation as the stimulation periods near successive multiples of the spontaneous period. The above results correspond to the model with parameters P^*=8 mv and T^*=50 ms. When P^* increases and T^* decreases, the interval around the discharge open to learning shrinks, which implies that the range of frequencies susceptible to being learned decreases. On the other hand, as P^* decreases and T^* increases, the behavior of this rule becomes more and more similar to that of the preceding one.

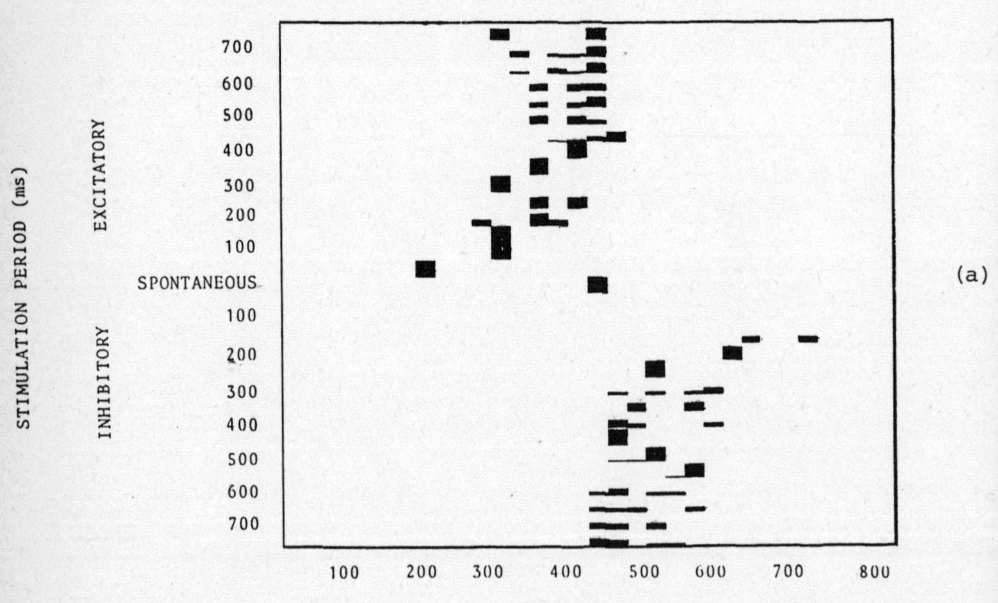

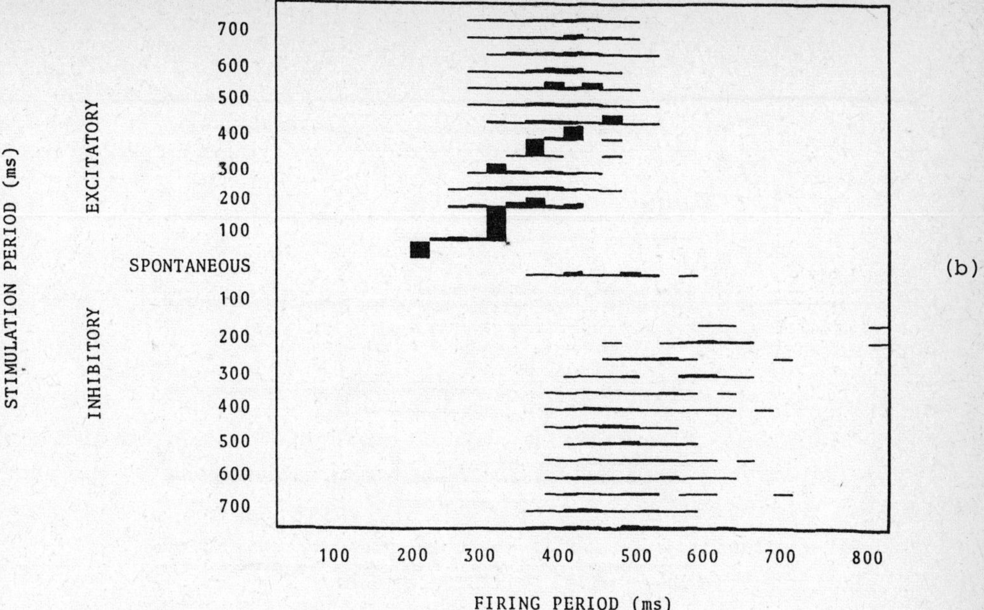

*** Figure 3.12 - Histograms of interspike intervals as a function of the period of a stimulus of amplitude 10 mv, for the model equipped with the third learning rule. (a) σ_{Pb_1}=0 mv. (b) σ_{Pb_1}=1 mv.

Period of the stimulus (ms)	Learned period (ms)	Period of the stimulus (ms)	Learned period (ms)	Period of the stimulus (ms)	Learned period (ms)
50	---	550	---	1050	---
100	---	600	---	1100	---
150	---	650	---	1150	---
200	---	700	---	1200	---
250	---	750	---	*	---
300	---	*	---	1250	415
350	---	800	400	*	425
*	---	*	412	1300	435
400	400	850	425	*	---
*	425	*	438	1350	---
450	450	900	---	1400	---
*	475	950	---	1450	---
500	---	1000	---	1500	---

*** Table 3.5 - Period learned by the neuron model, equipped with the third learning rule, as a function of the period of the stimulus presented.

Figure 3.13 shows the detailed evolution of the learning parameter m_{Pb_1} and of the output amplitude, for stimuli of periods between 150 and 700 ms. Since each point of the axis of abscissas corresponds to a discharge and the time elapsed is the same for all graphs, their respective

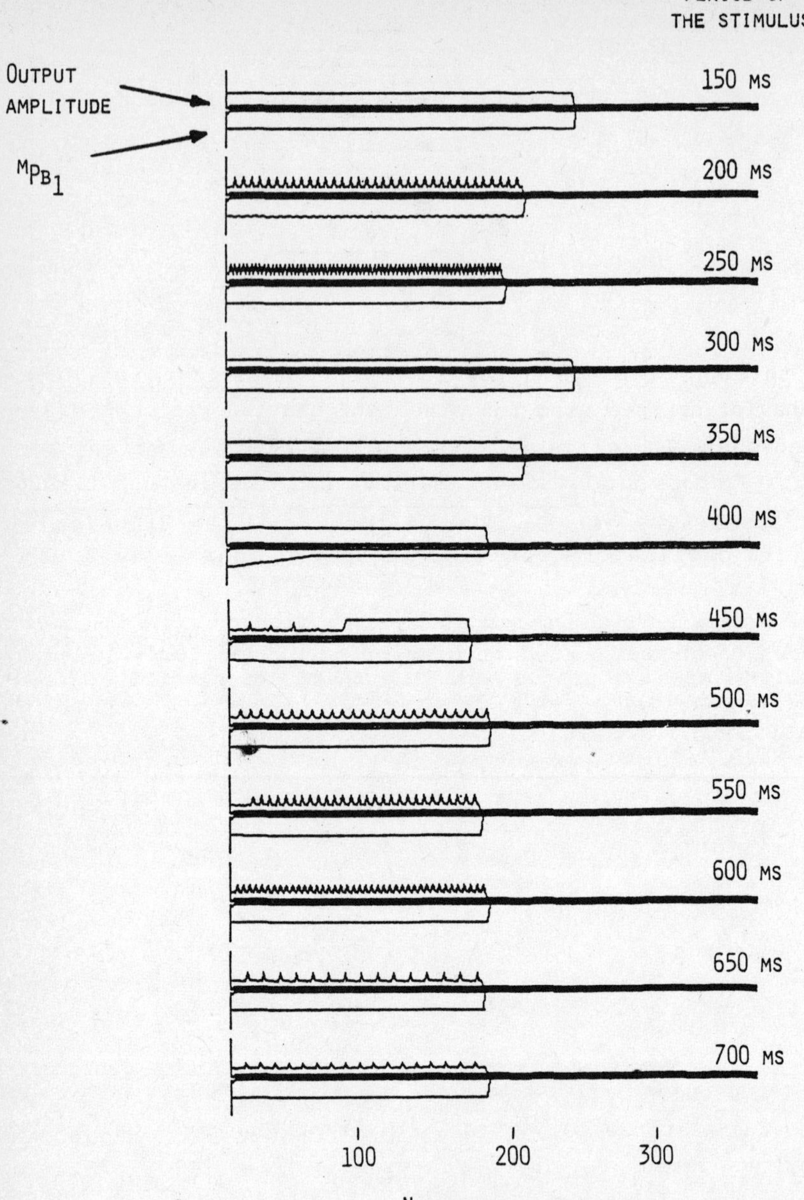

*** Figure 3.13 - Detailed evolution of the output amplitude (up) and of the learning parameter m_{Pb_1} (down), for stimuli of periods between 150 and 700 ms.

lengths are proportional to the neuron's firing frequency in each case. This frequency does not decrease monotonically as the period of the stimulus increases, but, rather, backward jumps provoked by entrainment take place, which are "paradoxical" in the sense described in the preceding section. Stationarity of m_{Pb_1} is clearly observed, except for the stimulation periods of 400 and 450 ms, in which the neuron learns the imposed rhythm, with m_{Pb_1} following a monotonically convergent trajectory. The oscillatory behavior of the parameter, in the second graph, derives from the fact that the stimulation impulses arrive successively just before and just after the discharge, nullifying in this way, through time, the momentary effects of learning. This same kind of behavior appears also for other stimulation periods between the represented ones. In relation to the output, it is important to point out that its average amplitude is maximum for the learned frequencies and that, in particular, for the period of 450 ms, increases with learning. Its oscillatory behavior in some graphs derives from the periodic variability of the phase of arrival of the stimulation impulse.

In summary, the proposed model equipped with this third rule satisfies in an acceptable way the theoretical and experimental requirements laid out in Section 3.1; therefore, the model thus equipped will be used in Chapter 5 as the basic node for the network.

3.4 - CONCLUSIONS

A pacemaker neuron model has been proposed, specified in terms of electrophysiological variables, that reproduces a set of phenomena observed in experimental preparations of pacemaker neurons in nervous systems of invertebrates.

The phenomena studied have been basically the ones relative to randomness and to the response to periodic trains of impulses: variability of the spontaneous interspike interval, phase-shifts provoked by occasional perturbations, (1:1) and superharmonic entrainment, and long-range changes in the spontaneous firing rate, caused by repeated rhythmic stimulation.

Specifically, the parameters of the model have been adjusted so as to reproduce the behavior of pacemaker neurons in the abdominal ganglion of Aplysia, according to the data provided by Perkel et al. (1964),

Junge and Moore (1966), Segundo and Perkel (1969), Kristan (1971), Parnas et al. (1974), Pinsker (1977a & b), Pinsker and Kandel (1977), Pinsker and Ayers (1983) and others. The probability density function of the spontaneous interspike interval has been derived analytically, while the phase response curve, the intervals of entrainment and the learned frequencies as a function of the stimulation frequencies have been obtained through simulation. The agreement between these results and experimental data is acceptable with regard to the first three aspects; in relation to learning, there is no organized experimental data on the subject, but the model behaves in a manner consistent with the indications that are found in the literature.

The ultimate objective of setting up a model of this kind was to have a basic node with which to build a network that would make it possible to study, at the microscopic level, certain phenomena of storage and recognition of temporal patterns of stimulation, observed at a macroscopic level. In this sense, the learning rule proposed conjectures a biologically plausible mechanism by means of which neurons, having access only to the magnitudes that characterize their instantaneous state, could process and record the frequency of the stimulation supplied to them.

Certain hypotheses on which the model is based and predictions derived from its simulation suggest experimental designs to corroborate or refute them. In relation to the first ones, only the impulses that leave a trace in the postsynaptic potential greater than a specific threshold at the time of firing have incidence upon accelerative learning, in the same way as only the impulses arriving just after discharge have incidence upon decelerative learning. The neuron could be stimulated with aperiodic trains of impulses, contingent upon one or the other condition, to observe if the neuron accelerates or decelerates its rhythm as conjectured.

Some predictions derived from the simulation results have already been commented on along the exposition, referring to the corresponding graphics. One that has not been made explicit is related to the limits of entrainment and of learning. The range of frequencies with which the neuron synchronizes widens considerably if, instead of stimulating it directly at a specific frequency, the spontaneous one is presented at the start and is varied progressively until it arrives at the desired frequency. In regard to entrainment, the above prediction has been confirmed experimentally in the stretch receptor of crayfish (Kohn et al., 1981). Not finding it reflected in the behavioral repertoires of the existing models, the need of a model that included it had already been suggested in the literature (Segundo and Kohn, 1981).

Apart from collateral contributions such as the one mentioned, the basic interest of the model resides in a new way of dealing with randomness and, above all, in the specification of a kind of learning that permits assimilation of imposed rhythms and that, being based on the temporal aspects of the stimulation, offers an alternative to the previously proposed learning rules.

CHAPTER 4
ANALYTIC STUDY OF THE ENTRAINMENT PATTERNS

"As theoreticians become more sophisticated in understanding neurophysiological findings and experimentalists become more sophisticated theoretically, their efforts will inevitably become mutually entrained".

H.M. Pinsker i J. Ayers, 1983

We start out, in the present chapter, from a simplified model of response to perturbations, consisting of a piecewise linear PRC, that approximates not only the one obtained for the model proposed in the preceding chapter, but also those obtained experimentally for a wide variety of pacemaker neurons with an excitatory synapse.

Up to the present, only the repertoire of phase transitions and the regions of subharmonic and superharmonic entrainment had been determined for this type of PRC (Segundo and Kohn, 1981), leaving as an open question the characterization of all other entrainment patterns, their regions of existence and their relative positions within the space of all possible stimulation conditions(£).

The prime objective of the chapter is to answer the above-stated question, characterizing the behavior of the pacemaker neuron submitted to all possible frequencies and amplitudes of stimulation. The procedure followed is: From the PRC, the phase transition equation is obtained, which is a difference equation on the torus. The regions where the different transitions take place are analyzed and the conditions which yield periodic solutions are determined, leading to the conclusion that the set of stimulation frequencies that cause entrainment is dense in the real line and that the ratio between the stimulation and firing

----

(£): "Important issues remain to be clarified: for example, the bounds of several ranges of entrainment, their relative positions, the λ (earlier EPSP arrival phase that triggers a spike) and E (presynaptic period) values for which they are meaningful, the presence or absence of a monotonicity with E of the slopes, etc. (...) It is unclear to us which ratios are swept on passing from the E with 1:r locking to that with 1:r+1 locking, as well as which are their order and corresponding E ranges" (Segundo and Kohn, 1981).

frequencies is a generalized Cantor function of the ratio between the stimulation and spontaneous frequencies. Next, a detailed characterization of the stimulus-neuron entrainment patterns that emerge is obtained, together with an effective procedure for computing them from the ratio between the stimulation and firing frequencies. Finally, the effect of randomness and learning upon the different entrainment patterns is analyzed and some avenues of future research are sketched.

Similar studies for other types of PRCs or for difference equations characterizing the behavior of a model, not necessarily in terms of phase transitions, have been carried out in three quite different contexts.

In the first place, the PRC obtained in experimental preparations of heart cells has been extensively analyzed, yielding a partial characterization of the input/output patterns that emerge (Ikeda et al., 1981; Ikeda, 1982) and of the regions of the parameter space where the different arhythmias originate (Guevara et al., 1981).

Secondly, the response of certain pacemaker neuron models --of the "integrate-and-fire" type or similar-- to continuous stimulation, either constant or periodic, is periodic. This has permitted deducing, from the differential equations that govern the evolution of the membrane potential, the functional dependence of the firing frequency on the stimulation amplitude (Nagumo and Sato, 1971; Yoshizawa et al., 1982) and on the stimulation frequency (Keener et al., 1981). Curiously enough, the curve that expresses the dependence on the stimulation amplitude presents topologic properties similar to those obtained, in the present chapter, for the dependence on the stimulation frequency in the case of a pacemaker neuron.

Finally, the conditions in which some types of rhythmic and arhythmic behaviors arise, in response to periodic stimuli, have been determined for several biological oscillator models: cardiac (Keener, 1981; Guevara and Glass, 1982), related to the breathing rhythm (Glass et al., 1980), circadian rhythm (Hoppensteadt and Keener, 1982) and generic rhythms (Glass and Mackey, 1979).

The effect of randomness upon these types of phenomena has been studied only in very particular cases (Knight, 1972; Glass et al., 1980) and almost always with the purpose of predicting the macroscopic behavior of a set of identical oscillators submitted to the same stimulation conditions.

4.1 - SIMPLIFIED MODEL OF RESPONSE TO PERIODIC STIMULI

In Chapter 2, the Phase Response Curve (PRC) has been defined and, in Chapter 3, those corresponding to several stimulation amplitudes have been computed through simulation, for the pacemaker neuron model proposed. In the present chapter, we will consider a piecewise linear approximation of that type of curve, defined as follows:

$$\delta_{\lambda}(\emptyset) = \begin{cases} -\left(\frac{1-\lambda}{\lambda}\right)\emptyset , & 0 \leq \emptyset < \lambda \\ -1+\emptyset , & \lambda \leq \emptyset < 1 \end{cases} \tag{4.1}$$

where $0<\lambda<1$ (Figure 4.1).

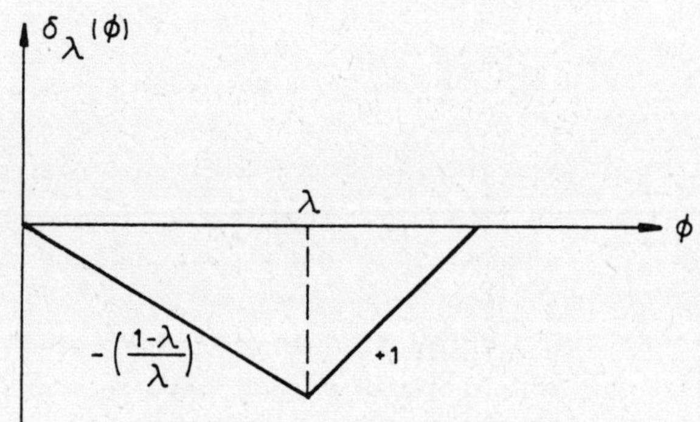

*** Figure 4.1 - Simplified model consisting of a piecewise linear PRC.

λ is the minimum phase for which the arrival of an input impulse triggers the immediate discharge of the neuron. λ decreases as the amplitude of the input impulse increases. Thus, leaving λ as a parameter, the PRC just defined can be thought to characterize the response of the pacemaker neuron to an input impulse of any amplitude.

The study of this type of PRC has a great interest from the experimental point of view, since it appears in many preparations (Segundo and Perkel, 1969; Hartline, 1976a; Ayers and Selverston, 1977a, 1979;

Beltz and Gelperin, 1980; Pinsker and Ayers, 1983; Buño and Fuentes, 1984, 1985).

In order to derive the Phase Advance Map from the PRC, we review some notation used in the preceding chapter and establish some new terminology as well.

T_0 : Period of the pacemaker neuron

T_1 : Period of the pulsed input

$$N = \frac{T_1}{T_0}$$

$\emptyset_n$: Phase of the pacemaker neuron at the time of arrival of the nth impulse

The **Phase Advance Map** --also called "Poincaré map"-- allows us to compute $\emptyset_{n+1}$ from $\emptyset_n$ (see Figure 4.2):

$$\emptyset_{n+1} = \langle \emptyset_n - \delta_\lambda(\emptyset_n) + N \rangle_1 \qquad (4.2)$$

where $\langle x \rangle_a$ stands for the remainder on dividing x by a.

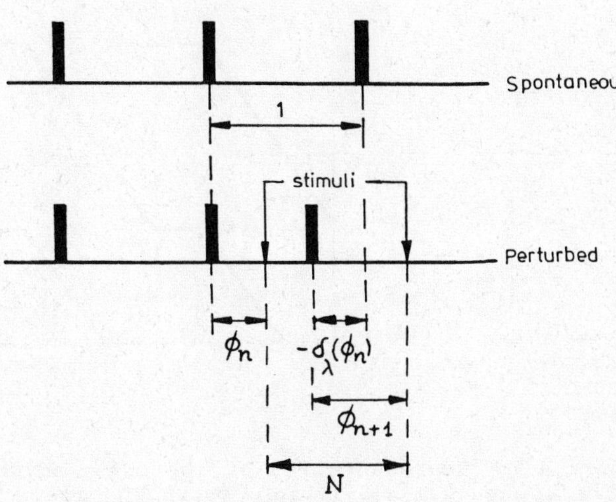

*** Figure 4.2 - Elements involved in the computation of the next phase of arrival of an input impulse ($\emptyset_{n+1}$) from the previous one ($\emptyset_n$).

Substituting δ_λ for expression 4.1 and taking into account that $\langle\langle x\rangle_a + \langle y\rangle_a\rangle_a = \langle x+y\rangle_a$:

$$\phi_{n+1} = g_{N,\lambda}(\phi_n) = \begin{cases} \langle \frac{1}{\lambda}\phi_n + N\rangle_1 , & 0 \leq \phi_n < \lambda \\ \langle N\rangle_1 , & \lambda \leq \phi_n < 1 \end{cases} \tag{4.3}$$

PROPOSITION 4.1. $g_{N,\lambda}$: S^1 ---> S^1 is a continuous onto mapping, which preserves the orientation of the circle and has continuous derivative except at $\phi=0$ and $\phi=\lambda$.

Proof - All the stated properties derive from the fact that:

$$\lim_{\phi\uparrow\lambda} g_{N,\lambda}(\phi) = \lim_{\phi\downarrow\lambda} g_{N,\lambda}(\phi) , \quad \lim_{\phi\uparrow 1} g_{N,\lambda}(\phi) = g_{N,\lambda}(0)$$

and from the corresponding properties of the functions defined in the intervals $[0,\lambda)$ and $[\lambda,1)$.∎

Consequently, $g_{N,\lambda}$ can be represented as a closed curve on the surface of a torus (Figure 4.3).

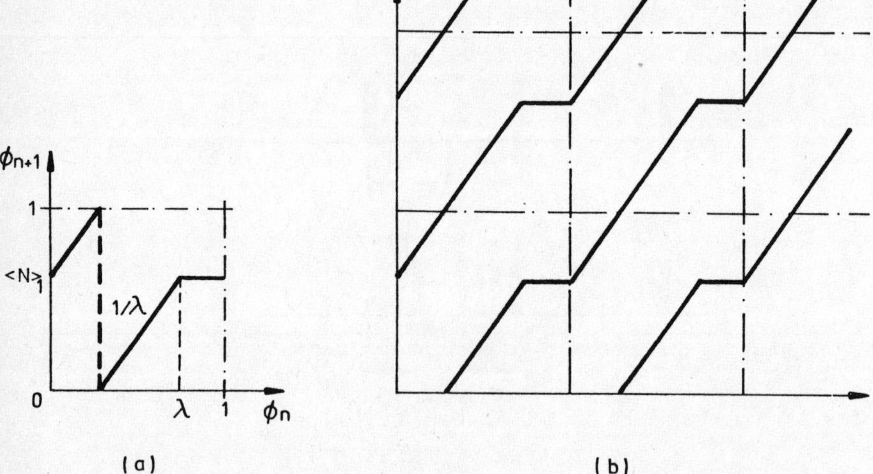

*** Figure 4.3 - Representation of the phase advance map: (a) in the space $[0,1)\times[0,1)$ and (b) in the space corresponding to the unfolding of the torus in $\mathbb{R}^2$.

Since: $\langle N_A\rangle_1 = \langle N_B\rangle_1 \Longrightarrow g_{N_B,\lambda} = g_{N_A,\lambda}$, it will suffice to study the behavior of equation 4.3 for $0\leq N<1$. From now on and until the contrary is explicitely stated, we will assume that N belongs to this interval and, therefore, for each λ, the phase advance map is considered to be defined on the torus.

With this restriction, we can redefine equation 4.3 in the following way:

$$\phi_{n+1} = g_{N,\lambda}(\phi_n) = \begin{cases} \frac{1}{\lambda}\phi_n + N \ , & 0 \leqslant \phi_n < \lambda\,(1-N) \quad \text{(a)} \\ \frac{1}{\lambda}\phi_n + N - 1 \ , & \lambda(1-N) \leqslant \phi_n < \lambda \quad \text{(ac)} \\ N \ , & \lambda \leqslant \phi_n < 1 \quad \text{(b)} \end{cases} \tag{4.4}$$

The labels attached to the branches of equation 4.4, when interpreted in the neural pacemaker context, correspond to the following events: (**a**) input impulse, (**ac**) input impulse followed by the discharge of the pacemaker, and (**b**) input impulse and simultaneous discharge of the pacemaker.

With the help of these labels, we define the **input/output pattern (I/O pattern)** of $g_{N,\lambda}$, for a given initial condition ϕ_1, as the string:

$$@_1 \ @_2 \ \cdots \ @_n \ \cdots$$

$$\text{where } @_i = \begin{cases} a & \text{if } \phi_i \varepsilon \ [0, \ \lambda(1-N)) \\ ac & \text{if } \phi_i \varepsilon \ [\lambda(1-N), \ \lambda) \\ b & \text{if } \phi_i \varepsilon \ [\lambda, \ 1) \end{cases}$$

Two remarks: First, when n **a**'s are followed by an **ac**, we will write $\mathbf{a}^{n+1}\mathbf{c}$ to abbreviate. Second, periodic patterns will be characterized by the string corresponding to one period.

4.2 - PHASE TRANSITION ANALYSIS

This section is devoted to the study of the conditions that determine the different transitions from each branch to another of equation 4.4. We will come up with a complete division of the space $[0,1)\times[0,1)$ --the Cartesian product of the initial phases ϕ_1 and the ratios between spontaneous and stimulation frequencies N-- into the regions where each transition takes place. The above division in regions permits deducing the type of I/O patterns that will appear, for each value of N.

Transition **a** ---> **a**

For an initial phase $\emptyset_1$ to determine the transition **a** ---> **a**, the following conditions must hold:

$$\begin{cases} 0 \leqslant \emptyset_1 < \lambda(1-N) \\ 0 \leqslant \frac{1}{\lambda}\emptyset_1 + N < \lambda(1-N) \end{cases}$$

Then:

$$\max(0, -\lambda N) \leqslant \emptyset_1 < \min(\lambda(1-N), \lambda(\lambda - \lambda N - N))$$

and taking into account that $0 \leqslant \lambda < 1$ and $0 \leqslant N < 1$:

$$0 \leqslant \emptyset_1 < \lambda(\lambda - \lambda N - N)$$

Consequently, this transition will only be possible if:

$$0 < \lambda(\lambda - \lambda N - N)$$

and isolating N:

$$0 \leqslant N < \frac{\lambda}{1+\lambda}$$

The initial conditions and regions of existence of the remaining transitions can be derived in the same way, so we will only enunciate them here.

Transition **a** ---> **ac**

$$\begin{cases} 0 \leqslant \emptyset_1 < \lambda(1-N) \\ \lambda(1-N) \leqslant \frac{1}{\lambda}\emptyset_1 + N < \lambda \end{cases}$$

$$\max(0, \lambda(\lambda - \lambda N - N)) \leqslant \emptyset_1 < \lambda(\lambda - N)$$

$$0 \leqslant N < \lambda$$

with:

$$\max(0, \lambda(\lambda-\lambda N-N)) = \begin{cases} \lambda(\lambda-\lambda N-N) & , \quad 0 \leqslant N < \frac{\lambda}{1+\lambda} \\ 0 & , \quad \frac{\lambda}{1+\lambda} \leqslant N < \lambda \end{cases}$$

<u>Transition a ---> b</u>

$$\begin{cases} 0 \leqslant \emptyset_1 < \lambda(1-N) \\ \lambda \leqslant \frac{1}{\lambda}\emptyset_1 + N < 1 \end{cases}$$

$$\max(0, \lambda(\lambda-N)) \leqslant \emptyset_1 < \lambda(1-N)$$

$$0 \leqslant N < 1$$

with:

$$\max(0, \lambda(\lambda-N)) = \begin{cases} \lambda(\lambda-N) & , \quad 0 \leqslant N < \lambda \\ 0 & , \quad \lambda \leqslant N < 1 \end{cases}$$

<u>Transition ac ---> a</u>

$$\begin{cases} \lambda(1-N) \leqslant \emptyset_1 < \lambda \\ 0 \leqslant \frac{1}{\lambda}\emptyset_1 + N - 1 < \lambda(1-N) \end{cases}$$

$$\lambda(1-N) \leqslant \emptyset_1 < \min(\lambda, \lambda(1+\lambda)(1-N))$$

$$0 \leqslant N < 1$$

with:

$$\min(\lambda, \lambda(1+\lambda)(1-N)) = \begin{cases} \lambda & , \quad 0 \leqslant N < \frac{\lambda}{1+\lambda} \\ \lambda(1+\lambda)(1-N) & , \quad \frac{\lambda}{1+\lambda} \leqslant N < 1 \end{cases}$$

<u>Transition ac ---> ac</u>

$$\begin{cases} \lambda(1-N) \leqslant \emptyset_1 < \lambda \\ \lambda(1-N) \leqslant \frac{1}{\lambda}\emptyset_1 + N - 1 < \lambda \end{cases}$$

$$\lambda(1+\lambda)(1-N) \leqslant \emptyset_1 < \min(\lambda, \lambda(1+\lambda-N))$$

$$\frac{\lambda}{1+\lambda} \leq N < 1$$

with:

$$\min(\lambda,\ \lambda(1+\lambda-N)) = \begin{cases} \lambda\ , & \frac{\lambda}{1+\lambda} \leq N < \lambda \\ \lambda(1+\lambda-N)\ , & \lambda \leq N < 1 \end{cases}$$

<u>Transition **ac ---> b**</u>

$$\begin{cases} \lambda(1-N) \leq \emptyset_1 < \lambda \\ \lambda \leq \frac{1}{\lambda}\emptyset_1 + N - 1 < 1 \end{cases}$$

$$\lambda(1+\lambda-N) \leq \emptyset_1 < \lambda$$

$$\lambda < N < 1$$

<u>Transition **b ---> a**</u>

$$\begin{cases} \lambda \leq \emptyset_1 < 1 \\ 0 \leq N < \lambda(1-N) \end{cases}$$

$$\lambda \leq \emptyset_1 < 1$$

$$0 \leq N < \frac{\lambda}{1+\lambda}$$

<u>Transition **b ---> ac**</u>

$$\begin{cases} \lambda \leq \emptyset_1 < 1 \\ \lambda(1-N) \leq N < \lambda \end{cases}$$

$$\lambda \leq \emptyset_1 < 1$$

$$\frac{\lambda}{1+\lambda} \leq N < \lambda$$

Transition b ---> b

$$\begin{cases} \lambda \leqslant \emptyset_1 < 1 \\ \lambda \leqslant N < 1 \end{cases}$$

Figure 4.4 synthesizes all the above results, showing the regions of the space of coordinates ($\emptyset_1$,N) where each transition originates. To provide a feeling for the diagram's evolution when varying λ , those corresponding to the values λ=0, 1/3, 2/3 and 1 are included. Note that $\lambda \epsilon (0,1)$ and, therefore, the extreme graphics are to be interpreted only as limits.

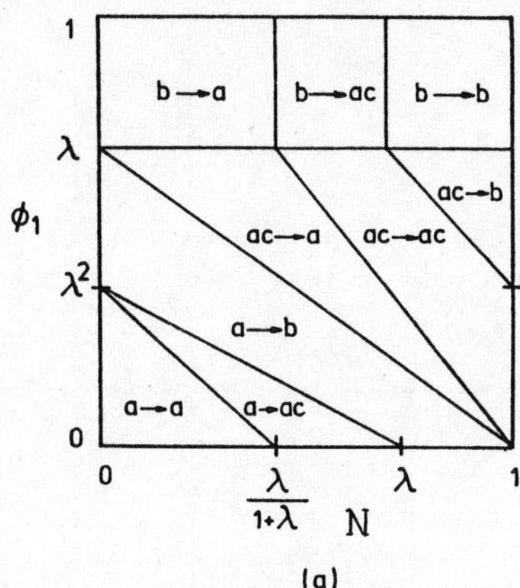

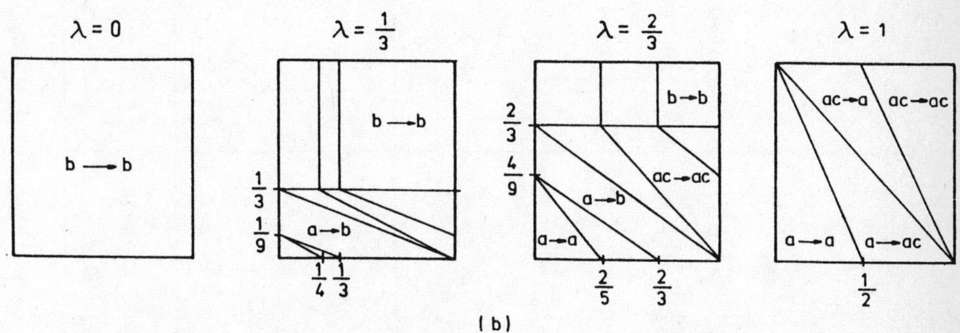

*** Figure 4.4 - Regions of the space of coordinates ($\emptyset_1$,N) where the different transitions originate: (a) general case, (b) cases λ = 0, 1/3, 2/3 and 1.

For each value of N, it is easy to derive from the above diagram the gamut of possible transitions, which implies a bare characterization of the type of I/O patterns that will appear (Table 4.1).

	$0 \leq N < \frac{\lambda}{1+\lambda}$	$\frac{\lambda}{1+\lambda} \leq N < \lambda$	$\lambda \leq N < 1$
TRANSITIONS THAT TAKE PLACE FOR EACH RANGE OF VALUES OF ϕ_1.	1 $b \to a$ λ $ac \to a$ $\lambda(1-N)$ $a \to b$ $\lambda(\lambda-N)$ $a \to ac$ $\lambda(\lambda-\lambda N-N)$ $a \to a$ 0	1 $b \to ac$ λ $ac \to ac$ $\lambda(1+\lambda)(1-N)$ $ac \to a$ $\lambda(1-N)$ $a \to b$ $\lambda(\lambda-N)$ $a \to ac$ 0	1 $b \to b$ λ $ac \to b$ $\lambda(1+\lambda-N)$ $ac \to ac$ $\lambda(1+\lambda)(1-N)$ $ac \to a$ $\lambda(1-N)$ $a \to b$ 0
POSSIBLE TRANSITIONS	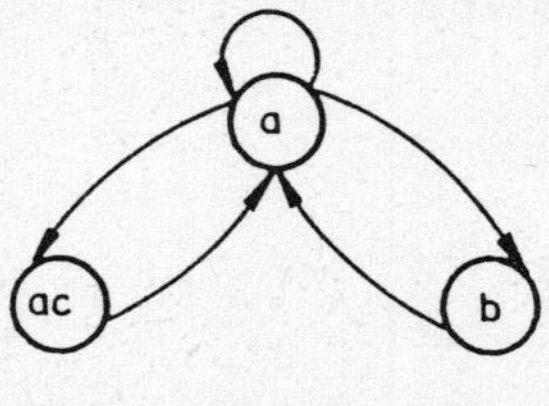	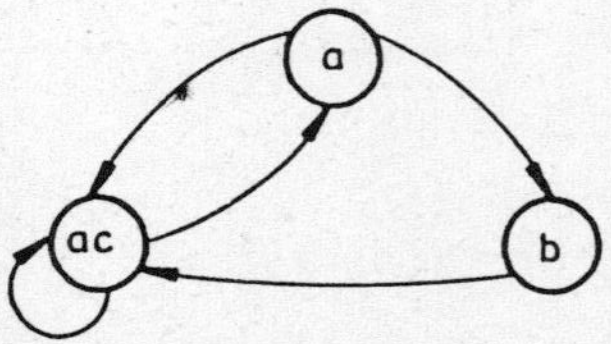	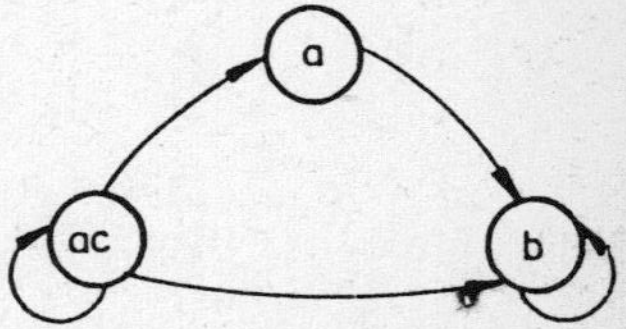
INPUT/OUTPUT PATTERN $@_0\ @_1\ \ldots\ @_i\ \ldots$	$@_i = b \vee @_i = ac \Rightarrow$ $\Rightarrow @_{i-1} = @_{i+1} = a$	$@_i = a \Rightarrow @_{i-1} = ac$ $@_i = b \Rightarrow @_{i-1} = a \wedge @_{i+1} = ac$	$@_i = a \Rightarrow @_{i-1} = ac \wedge @_{i+1} = b$ $@_i = b \Rightarrow @_k = b,\ \forall k > i$

*** Table 4.1 - Possible transitions and restrictions upon the I/O patterns that appear, for the different intervals of values of N.

To refine this characterization, it is necessary to study the periodic solutions of equation 4.4, as we will do in the next section.

The phase transition analysis carried out for $N \in [0,1)$ can be easily extended to $N \in \mathbb{R}^+$, by just substituting N by $<N>_1$ in the boundaries of the regions where each transition originates and placing $(N-<N>_1)$ symbols **c** before each label **a, b** and **ac.**

4.3 - PERIODIC SOLUTIONS

The solutions of period n of equation 4.4 are the fixed points of the mapping $g^n_{N,\lambda}$ that are not left invariant by the mappings $g^m_{N,\lambda}$, for $m<n$.

4.3.1 - Solutions of period 1

"Practise yourself, for heaven's sake, in little things; and thence proceed to greater".

Epictetus

The objective is to find the fixed points of the mapping $g_{N,\lambda}$, corresponding to the points where the diagonal $\emptyset_{n+1} = \emptyset_n$ crosses the curve $\emptyset_{n+1} = g_{N,\lambda}(\emptyset_n)$ in Figure 4.3.

PROPOSITION 4.2. The only solutions of equation $g_{N,\lambda}(\emptyset) = \emptyset$ are:

$$\emptyset^* = 0 \quad \text{for} \quad N = 0\ ;$$

$$\emptyset^* = \lambda \quad \text{for} \quad N = \lambda\ ; \quad \text{and}$$

$$\emptyset^{*(1)} = \frac{\lambda(1-N)}{1-\lambda} \ \wedge\ \emptyset^{*(2)} = N \quad , \quad \forall N \ \varepsilon\ (\lambda,\ 1)$$

Proof - Because of expression 4.4, $\emptyset^*$ is a solution of equation $g_{N,\lambda}(\emptyset) = \emptyset$ iff:

$$\emptyset^* = \frac{1}{\lambda}\,\emptyset^* + N \quad \text{with} \quad 0 \leqslant \emptyset^* < \lambda(1-N)$$

$$\text{or} \quad \emptyset^* = \frac{1}{\lambda}\,\emptyset^* + N-1 \quad \text{with} \quad \lambda(1-N) \leqslant \emptyset^* < \lambda$$

$$\text{or} \quad \emptyset^* = N \quad \text{with} \quad \lambda \leqslant \emptyset^* < 1$$

From the first condition we have:

$$\varnothing^* = N \frac{\lambda}{\lambda-1}$$

Since $0 \leqslant \lambda < 1$:

$$\frac{\lambda}{\lambda-1} < 0$$

and taking into account that $N \geqslant 0$, it results that the condition is only satisfied for $\varnothing^*=0$ and $N=0$.

From the second condition we deduce:

$$\varnothing^* = \frac{\lambda(1-N)}{1-\lambda} \quad \text{with} \quad \lambda(1-N) \leqslant \frac{\lambda(1-N)}{1-\lambda} < \lambda$$

which is the same as: $N > \lambda$.

From the third condition we have: $\varnothing^*=N$ with $N \geqslant \lambda$. ▮

PROPOSITION 4.3. $\varnothing^*=0$ for $N=0$, $\varnothing^*=\lambda$ for $N=\lambda$ and

$\varnothing^{*(1)} = \dfrac{\lambda(1-N)}{1-\lambda}$, $\forall N \in (\lambda, 1)$ are unstable solutions.

$\varnothing^{*(2)} = N$, $\forall N \in (\lambda, 1)$ is a stable solution.

Proof - Because of a well-known result of the theory of difference equations (Bernussou, 1977), a solution $\varnothing^*$ of $g_{N,\lambda}(\varnothing) = \varnothing$ is stable iff:

$$\left| \left[\frac{dg_{N,\lambda}(\varnothing)}{d\varnothing} \right]_{\varnothing = \varnothing^*} \right| < 1.$$

But:

$$\left[\frac{dg_{N,\lambda}(\emptyset)}{d\emptyset}\right]_{\emptyset = \emptyset^*} = \frac{1}{\lambda} > 1 \ , \ \forall\emptyset^* \in (0,\lambda)$$

$$\left[\frac{dg_{N,\lambda}(\emptyset)}{d\emptyset}\right]_{\emptyset = \emptyset^*} = 0 \ , \ \forall\emptyset^* \in (\lambda, 1) \qquad (4.5)$$

and $g_{N,\lambda}$ is not derivable for N=0 and N=λ. ▮

In neural terms, an entrainment pattern is stable if it emerges again after having suffered a transitory perturbation; while it is unstable if any occasional variation of the stimulation conditions provokes its definitive disappearance.

The solutions of period 1, for $N \in [0,1)$, correspond to one-to-one entrainment and, for $N \in \mathbb{R}^+$, correspond to (1:r) superharmonic entrainments or, in other words, those in which a stimulation impulse arrives every r pacemaker discharges ($r = N_A - \langle N_A \rangle_1$). In the unstable solution **--ac--**, the stimulation impulse does not occur simultaneously with any pacemaker discharge. In the stable solution **--b--**, the stimulation impulse always occurs simultaneously with a pacemaker discharge.

4.3.2 - Solutions of period 2

From the phase transition study carried out in Section 4.2, we can derive the equation that defines $g^2_{N,\lambda}$:

$$\emptyset_{n+2} = g^2_{N,\lambda}(\emptyset_n) = \begin{cases} \frac{1}{\lambda^2}\emptyset_n + \frac{1}{\lambda}N + N, & 0 \leqslant \emptyset_n < \lambda(\lambda-\lambda N-N) \quad [a \longrightarrow a] \\ \frac{1}{\lambda^2}\emptyset_n + \frac{1}{\lambda}N + N - 1, & \lambda(\lambda-\lambda N-N) \leqslant \emptyset_n < \lambda(\lambda-N) \quad [a \longrightarrow ac] \\ N', & \lambda(\lambda-N) \leqslant \emptyset_n < \lambda(1-N) \quad [a \longrightarrow b] \\ \frac{1}{\lambda^2}\emptyset_n + \frac{1+\lambda}{\lambda}N - \frac{1}{\lambda}, & \lambda(1-N) \leqslant \emptyset_n < \\ & < \min(\lambda, \lambda(1+\lambda)(1-N)) \quad [ac \longrightarrow a] \\ \frac{1}{\lambda^2}\emptyset_n + \frac{1+\lambda}{\lambda}N - \frac{1+\lambda}{\lambda}, & \lambda(1+\lambda)(1-N) \leqslant \emptyset_n < \\ & < \min(\lambda, \lambda(1+\lambda-N)) \quad [ac \longrightarrow ac] \\ N, & \lambda(1+\lambda-N) \leqslant \emptyset_n < \lambda \quad [ac \longrightarrow b] \\ \frac{1+\lambda}{\lambda}N, & \lambda \leqslant \emptyset_n < 1 \wedge 0 \leqslant N < \frac{\lambda}{1+\lambda} \quad [b \longrightarrow a] \\ \frac{1+\lambda}{\lambda}N - 1, & \lambda \leqslant \emptyset_n < 1 \wedge \frac{\lambda}{1+\lambda} \leqslant N < \lambda \quad [b \longrightarrow ac] \\ N, & \lambda \leqslant \emptyset_n < 1 \wedge \lambda \leqslant N < 1 \quad [b \longrightarrow b] \end{cases} \tag{4.6}$$

Figure 4.5 shows the shape taken by this function in each of the three intervals of values of N that have already been dealt with separately in Table 4.1. The solutions of equation $g^2_{N,\lambda}(\emptyset) = \emptyset$ correspond to the intersection points of the diagonal $\emptyset_{n+2} = \emptyset_n$ with the curves shown in this figure.

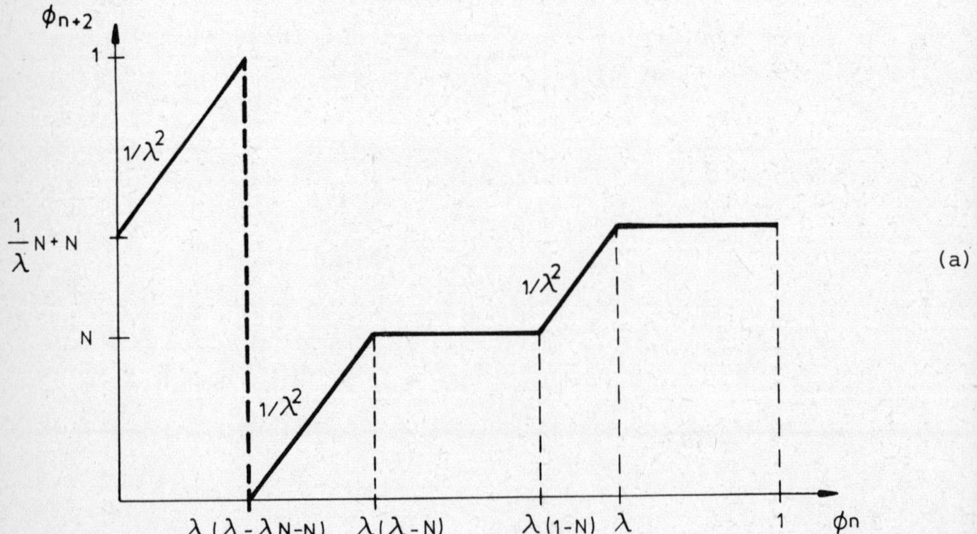

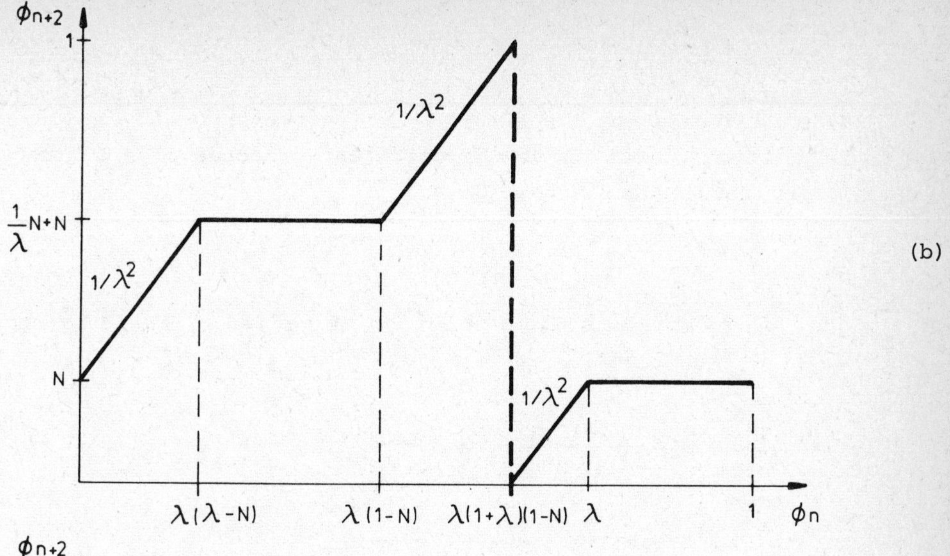

(b)

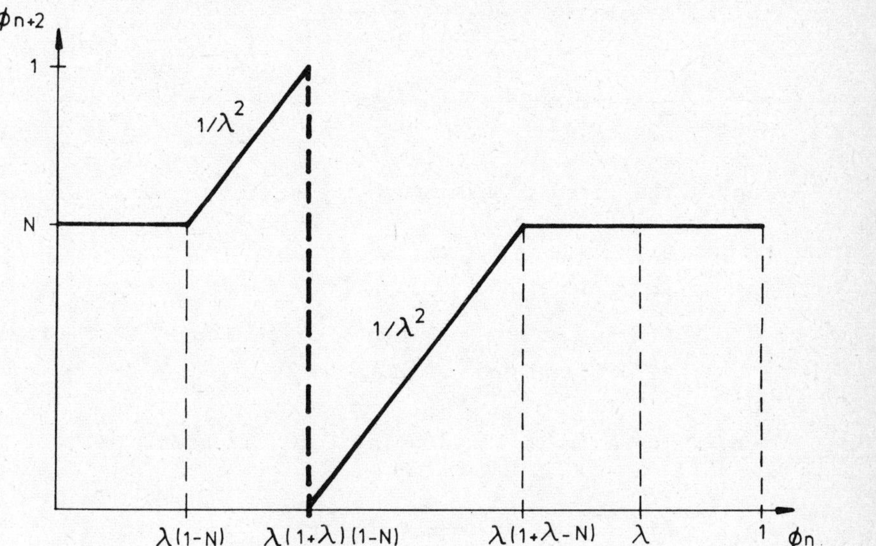

(c)

*** Figure 4.5 - Graphical representation of the function $g^2_{N,\lambda}$ for three intervals of values of N: (a) $0<N<\frac{\lambda}{1+\lambda}$; (b) $\frac{\lambda}{1+\lambda}<N<\lambda$; (c) $\lambda<N<1$.

PROPOSITION 4.4. The only solutions of equation $g^2_{N,\lambda}(\emptyset)=\emptyset$ are:

$$\emptyset^{*(1)} = N \quad \text{and} \quad \emptyset^{*(2)} = \lambda , \quad \text{for} \quad N = \frac{\lambda^2}{1+\lambda}$$

$$\left.\begin{array}{l} \emptyset^{*(1)} = \dfrac{\lambda(\lambda-\lambda N-N)}{1-\lambda^2} , \quad \emptyset^{*(2)} = N \\[2ex] \emptyset^{*(3)} = \dfrac{\lambda(1-\lambda N-N)}{1-\lambda^2} , \quad \emptyset^{*(4)} = \dfrac{1+\lambda}{\lambda}N \end{array}\right\}, \ \forall N \ \varepsilon \ \left(\frac{\lambda^2}{1+\lambda}, \frac{\lambda}{1+\lambda}\right)$$

$$\emptyset^{*(1)} = 0 \quad \text{and} \quad \emptyset^{*(2)} = \frac{\lambda}{1+\lambda}, \text{ for } \quad N = \frac{\lambda}{1+\lambda}$$

Proof - We will proceed the same way as in the proof of Proposition 4.2. Through the application of the different branches of equation 4.6, the following results are obtained:

[a ---> a] $\emptyset^* = \dfrac{N\lambda}{\lambda-1}$. The only fixed point is $\emptyset^*=0$ for $N=0$, which is not a solution of period 2, since by Proposition 4.2 it is also a fixed point of $g_{N,\lambda}$.

[a ---> ac] $\emptyset^* = \dfrac{\lambda(\lambda-\lambda N-N)}{1-\lambda^2}$ for $\dfrac{\lambda^2}{1+\lambda} < N \leqslant \dfrac{\lambda}{1+\lambda}$

[a ---> b] $\emptyset^* = N$ for $\dfrac{\lambda^2}{1+\lambda} \leqslant N < \dfrac{\lambda}{1+\lambda}$

[ac ---> a] $\emptyset^* = \dfrac{\lambda(1-\lambda N-N)}{1-\lambda^2}$ for $\dfrac{\lambda^2}{1+\lambda} < N \leqslant \dfrac{\lambda}{1+\lambda}$

[ac ---> ac] The fixed points are:

$$\emptyset^* = \frac{\lambda(1-N)}{1-\lambda}, \quad \forall N > \lambda$$

which are not solutions of period 2, since they are also fixed points of $g_{N,\lambda}$.

[ac ---> b] There is no fixed point for this branch.

[b ---> a] $\emptyset^* = \dfrac{1+\lambda}{\lambda} N$ for $\dfrac{\lambda^2}{1+\lambda} \leqslant N < \dfrac{\lambda}{1+\lambda}$

[b ---> ac] There is no fixed point for this branch.

[b ---> b] The fixed points are:

$$\emptyset^* = N, \ \forall N \in [\lambda, 1)$$

which are not solutions of period 2, since they are also fixed points of $g_{N,\lambda}$. ∎

PROPOSITION 4.5. All solutions of equation $g^2_{N,\lambda}(\emptyset)=\emptyset$ are unstable, except:

$$\emptyset^{*(2)} = N \wedge \emptyset^{*(4)} = \frac{1+\lambda}{\lambda} N , \quad \forall N \in (\frac{\lambda^2}{1+\lambda}, \frac{\lambda}{1+\lambda})$$

Proof - This follows from the application of the multidimensional version of the result used in the proof of Proposition 4.3, which establishes as a necessary and sufficient condition for $\emptyset^*$ to be a stable solution of equation $g^2_{N,\lambda}(\emptyset)=\emptyset$ (Bernussou, 1977) that:

$$\left| \left[\frac{\partial g_{N,\lambda}(\emptyset)}{\partial \emptyset}\right]_{\emptyset=\emptyset^*} \cdot \left[\frac{\partial g_{N,\lambda}(\emptyset)}{\partial \emptyset}\right]_{\emptyset = g_{N,\lambda}(\emptyset^*)} \right| < 1$$

Because of equations 4.5, the above inequality only holds when:

$$\emptyset^* \in (\lambda,1) \quad \text{or} \quad g_{N,\lambda}(\emptyset^*) \in (\lambda,1)$$

For $N=\frac{\lambda^2}{1+\lambda}$,

$$\emptyset^{*(1)} = g_{N,\lambda}(\emptyset^{*(2)}) = N = \frac{\lambda^2}{1+\lambda} < \lambda$$

$$\emptyset^{*(2)} = g_{N,\lambda}(\emptyset^{*(1)}) = \lambda$$

For $N \in (\frac{\lambda^2}{1+\lambda}, \frac{\lambda}{1+\lambda})$,

$$\emptyset^{*(1)} = g_{N,\lambda}(\emptyset^{*(3)}) = \frac{\lambda(\lambda-\lambda N-N)}{1-\lambda^2} < \frac{\lambda(\lambda-(1+\lambda)(\frac{\lambda^2}{1+\lambda}))}{1-\lambda^2} < \lambda$$

$$\emptyset^{*(2)} = g_{N,\lambda}(\emptyset^{*(4)}) = N < \frac{\lambda}{1+\lambda} < \lambda$$

$$\emptyset^{*(3)} = g_{N,\lambda}(\emptyset^{*(1)}) = \frac{\lambda(1-\lambda N-N)}{1-\lambda^2} < \frac{\lambda(1-(1+\lambda)(\frac{\lambda^2}{1+\lambda}))}{1-\lambda^2} = \lambda$$

$$\emptyset^{*(4)} = g_{N,\lambda}(\emptyset^{*(2)}) = \frac{1+\lambda}{\lambda}N > \frac{1+\lambda}{\lambda} \cdot \frac{\lambda^2}{1+\lambda} = \lambda$$

For $N=\frac{\lambda}{1+\lambda}$,

$$\phi^{*(1)} = g_{N,\lambda}(\phi^{*(2)}) = 0$$

$$\phi^{*(2)} = g_{N,\lambda}(\phi^{*(1)}) = N = \frac{\lambda}{1+\lambda} < \lambda$$

In sum, only $\phi^{*(4)}$ and $g_{N,\lambda}(\phi^{*(2)})$ belong to the interval $(\lambda,1)$, which implies that the only stable solutions are $\phi^{*(2)}$ and $\phi^{*(4)}$.∎

The solutions $\phi^{*(1)}$ and $\phi^{*(3)}$ actually correspond to the same periodic output, except that they are described starting at different instants in the cycle: **aac** for the former, and **aca** for the latter. The same occurs with $\phi^{*(2)}$ and $\phi^{*(4)}$: **ab** for the former, and **ba** for the latter. In both solutions, the stimulation frequency is double the firing frequency.

4.3.3 - Stable solutions of period n

In this section we will prove a series of general results characterizing the entrainment behavior of the simplified model. Because of the high level of technicality involved in some of the proofs, we give, at the beginning, an informal account of the results and their implications, so that the reader not interested in the mathematical details can skip them and go directly to Section 4.4.

Our first result is a characterization of stable periodic solutions as those that go through branch (b) of equation 4.4. Since this branch leads to a fixed subsequent phase $\phi=N$, there will be only one **b** in each cycle of a periodic I/O pattern. To simplify later computations, we adopt the convention of **b** marking the end of period, so that the initial phase will always be $\phi_1=N$. Experimentally, this convention would mean taking the arrival of an input impulse simultaneous with a discharge as the mark of end of period.

Next, we prove that there is only one I/O pattern for each entrainment ratio. This is important because we are guaranteed that, even under stimulation of different amplitudes and frequencies, if we get the same stimulus-neuron entrainment ratio, the detailed I/O pattern will be the same.

The previous result leads to the determination of the boundaries of all entrainment regions, which in turn permits proving that the ratio between the stimulation and firing frequencies is a generalized Cantor function of the ratio between the stimulation and spontaneous frequencies. Two consequences are worth mentioning: First, all stimulation frequencies, but a set of measure zero, give place to entrainment. Second, each entrainment ratio is originated by a set of input frequencies of positive measure. Thus, if the ideal conditions assumed by the simplified model held, entrainment would always arise when submitting the type of pacemakers considered to rhythmic stimulation and, for each entrainment ratio, we could find a whole band of stimulation frequencies yielding it.

By stratifying the rational numbers contained in the interval $[0,1)$ in infinite levels, we come up with an effective procedure to compute the I/O patterns with entrainment ratio in a given level from the I/O patterns with entrainment ratios in the preceding level. In this way, we provide a recursive characterization of all entrainment patterns, as well as an algorithm to compute any one of them from its entrainment ratio. This complete characterization of the repertoire of I/O patterns, for the deterministic simplified model, permits analyzing the effect of randomness and learning upon these patterns, as we will indicate in Section 4.4, thus contributing to elucidate if the mentioned two sources of divergence from the deterministic model are enough to account for the differences observed between the computed and the experimental patterns.

From this point and until the end of the section, we will provide mathematical proofs for the results just stated.

Our study will be limited to the solutions which are stable, since in a milieu so prone to perturbations as is the biological, they are the only ones liable to appear.

PROPOSITION 4.6. A solution $\emptyset^*$ of equation $g^n_{N,\lambda}(\emptyset^*)=\emptyset^*$ is stable iff there exists $m \leq n$ s.t. $\lambda < g^m_{N,\lambda}(\emptyset^*) < 1$.

Proof - For the solution $\emptyset^*$ to be stable, it has to satisfy (Bernussou, 1977):

$$\left| \prod_{i=0}^{n-1} \left[\frac{\partial g_{N,\lambda}(\emptyset)}{\partial \emptyset} \right]_{\emptyset = g^{i}_{N,\lambda}(\emptyset^*)} \right| < 1 \qquad (4.7)$$

But,

$$\left[\frac{\partial g_{N,\lambda}(\emptyset)}{\partial \emptyset} \right]_{\emptyset = \emptyset'} > 1 \ , \quad \forall \emptyset' \in (0, \lambda)$$

$$\left[\frac{\partial g_{N,\lambda}(\emptyset)}{\partial \emptyset} \right]_{\emptyset = \emptyset'} = 0 \ , \quad \forall \emptyset' \in (\lambda, 1)$$

and $g_{N,\lambda}$ is not derivable for $\emptyset=0$ and $\emptyset=\lambda$.

Therefore, for equation 4.7 to be satisfied, there must exist $m \leqslant n$ s.t. $g^{m}_{N,\lambda}(\emptyset^*) \in (\lambda,1)$.∎

Thus, a stable periodic solution has necessarily to go through branch (b) of equation 4.4 once and only once in each cycle, since this event creates a fixed condition $\emptyset$=N. We can then choose a representative of all solutions with the same periodic output and that differ only in their starting points in the cycle.

We will choose as representatives the periodic solutions that finish in branch (b), that is, those with $\emptyset_1$=N. All I/O patterns corresponding to entrainment situations will thus end up with the only **b** in the string.

Applying repeatedly the branches (a) and (ac) of equation 4.4, for the initial condition $\emptyset_1$=N, we obtain:

$$\emptyset_n = g^{n-1}_{N,\lambda}(N) = \sum_{i=0}^{n-1} \frac{1}{\lambda^i} N - \sum_{i=1}^{n-1} \frac{1}{\lambda^{n-i-1}} \mu_i$$

$$\text{with} \quad \mu_i = \begin{cases} 0 \text{ if } \emptyset_i \in [0,\lambda(1-N)) & \text{(a)} \\ 1 \text{ if } \emptyset_i \in [\lambda(1-N),\lambda) & \text{(ac)} \end{cases} \qquad (4.8)$$

(s:1) subharmonic entrainment takes place when the neuron fires every **s** stimulation impulses; in other words, it corresponds to solutions of the form $\mathbf{a}^{s-1}\mathbf{b}$.

PROPOSITION 4.7. The region of existence of stable (s:1) subharmonic entrainment is:

$$\frac{\lambda^s}{\sum_{i=0}^{s-1} \lambda^i} < N < \frac{\lambda^{s-1}}{\sum_{i=0}^{s-1} \lambda^i}$$

Proof - The conditions that determine the existence of stable solutions of the type (s:1) are:

$$\begin{cases} 0 \leqslant \emptyset_k < \lambda(1-N), & \forall k < s \\ \lambda < \emptyset_s < 1 \end{cases}$$

Using equation 4.8 and taking into account that in this case $\mu_i=0$, $\forall i<s$, there results:

$$\begin{cases} 0 \leqslant \sum_{i=0}^{k-1} \frac{1}{\lambda^i} N < \lambda(1-N), & \forall k < s \\ \lambda < \sum_{i=0}^{s-1} \frac{1}{\lambda^i} N < 1 \end{cases}$$

from which the stated region of existence is deduced.∎

(s:r) entrainment takes place when the neuron fires **r** times for every **s** input impulses.

For the study of this more generic type of entrainment, we will use some properties of a kind of mapping from the unit circle onto itself (see appendix B).

4.3.3.1 - Application of some properties of mappings from the unit circle onto itself to the function $g_{N,\lambda}$.

"Y llegaron al convencimiento de que 'la naturaleza está escrita en caracteres matemáticos', cuando lo que estaba escrito en caracteres matemáticos no era la naturaleza, sino... la estructura matemática de la naturaleza".

E. Sábato, 1951

We can define, as in appendix B, the mapping $\hat{g}_{N,\lambda}$: $\mathbb{R}$ ---> $\mathbb{R}$ induced from $g_{N,\lambda}$: S^1 ---> S^1 by the covering of $\mathbb{R}$ by S^1:

$$\hat{\phi}_{n+1} = \hat{g}_{N,\lambda}(\hat{\phi}_n) = \begin{cases} \hat{\phi}_n + \left(\frac{1-\lambda}{\lambda}\right) \langle\hat{\phi}_n\rangle_1 + N, & 0 \leqslant \langle\hat{\phi}_n\rangle_1 < \lambda \quad \text{(a)(ac)} \\ \hat{\phi}_n - \langle\hat{\phi}_n\rangle_1 + N + 1, & \lambda \leqslant \langle\hat{\phi}_n\rangle_1 < 1 \quad \text{(b)} \end{cases} \tag{4.9}$$

and iterating this equation for $\hat{\phi}_1$=N:

$$\hat{\phi}_n = \hat{g}^{n-1}_{N,\lambda}(N) = \sum_{i=0}^{n-1} \frac{1}{\lambda^i} N - (1-\lambda) \sum_{i=1}^{n-2} \frac{1}{\lambda^{n-i-1}} \sum_{j=1}^{i} \mu_j ,$$

$$\mu_j = \begin{cases} 0 \text{ if } \langle\hat{\phi}_j\rangle_1 \in [0,\lambda(1-N)) & \text{(a)} \\ 1 \text{ if } \langle\hat{\phi}_j\rangle_1 \in [\lambda(1-N),\lambda) & \text{(ac)} \end{cases} \tag{4.10}$$

Because of Propositions 1 and B.1, $\hat{g}_{N,\lambda}$ is a monotonically increasing, continuous onto mapping and, consequently, so are the mappings resulting from its iteration $\hat{g}^n_{N,\lambda}$, $\forall n \in \mathbb{N}$.

The **rotation number** of $g_{N,\lambda}$ is:

$$\rho_{N,\lambda} = \lim_{n\to\infty} \frac{\hat{\phi}_n}{n}$$

In neural terms, $\rho_{N,\lambda}$ represents the average number of pacemaker discharges per stimulation impulse.

PROPOSITION 4.8. $\rho_{N,\lambda}=(r/s)$ with $r,s\in\mathbb{N}$, $s\neq 0$ and $(r,s)=1 \Longrightarrow$ the stable I/O pattern of $g_{N,\lambda}$ contains (s-r) **a**'s, (r-1) **ac**'s and one **b**.

Proof - From Proposition B.3, the aforementioned I/O pattern will be periodic and, because of the convention adopted in relation to periodic I/O patterns, will end with a **b**.

Applying Proposition B.4 to branch (b) of equation 4.9 for $\emptyset^*=N$:

$$\hat{\emptyset}_s - \langle\hat{\emptyset}_s\rangle_1 + N + 1 = N + r$$

thus,

$$\hat{\emptyset}_s - \langle\hat{\emptyset}_s\rangle_1 = r-1 \tag{4.11}$$

All previous $\hat{\emptyset}_i$ will require the application of the other branch of equation 4.9 and, consequently, $\forall i=1,2,\ldots s-1$:

$$\left[\hat{\emptyset}_{i+1} - \langle\hat{\emptyset}_{i+1}\rangle_1\right] - \left[\hat{\emptyset}_i - \langle\hat{\emptyset}_i\rangle_1\right] =$$

$$= \left[\hat{\emptyset}_i + \left(\frac{1-\lambda}{\lambda}\right)\langle\hat{\emptyset}_i\rangle_1 + N - \langle\hat{\emptyset}_i + \left(\frac{1-\lambda}{\lambda}\right)\langle\hat{\emptyset}_i\rangle_1 + N\rangle_1\right] - \left[\hat{\emptyset}_i - \langle\hat{\emptyset}_i\rangle_1\right] =$$

$$= \frac{1}{\lambda}\langle\hat{\emptyset}_i\rangle_1 + N - \langle\frac{1}{\lambda}\langle\hat{\emptyset}_i\rangle_1 + N\rangle_1 = \begin{cases} 0 & \text{if } \langle\hat{\emptyset}_i\rangle_1 \in [0,\lambda(1-N)) \quad \text{(a)} \\ 1 & \text{if } \langle\hat{\emptyset}_i\rangle_1 \in [\lambda(1-N),\lambda) \quad \text{(ac)} \end{cases}$$

Therefore, there must be (r-1) **ac**'s in the I/O pattern. The remaining (s-r) symbols will be **a**'s. ∎

The case $\rho_{N,\lambda}=(r/s)$, with r/s an irreducible fraction, corresponds then to (s:r) entrainment.

Since the mapping $g_{N,\lambda}$ depends on the parameter $N\in[0,1)$, we can also define as in appendix B:

$$g_\lambda : [0,1) \times S^1 \longrightarrow S^1$$
$$(N, x) \longmapsto g_{N,\lambda}(x)$$

$$\hat{g}_\lambda \; : \; [0,1) \times \mathbb{R} \longrightarrow \mathbb{R}$$
$$(N, x) \longmapsto \hat{g}_{N,\lambda}(x)$$

$$\rho_{g_\lambda} \; : \; [0,1) \longrightarrow \mathbb{R}$$
$$N \longmapsto \rho_{N,\lambda}$$

From the definition of $\hat{g}_{N,\lambda}$, given in equation 4.9, we deduce that the mapping $\hat{g}_\lambda$ is continuous and monotonically increasing with N; hence, by Proposition B.5, so is ρ_{g_λ}.

4.3.3.2 - Regions of stable entrainment within the (N,λ)-parameter space

We will prove in the present section that ρ_{g_λ} is a generalized Cantor function and therefore, for a given λ, the set of values of N which yield an irrational $\rho_{N,\lambda}$ is of measure zero.

In neural terms, stable entrainment between a pacemaker and a rhythmic stimulus will always occur, except for a set of measure zero of input frequencies.

PROPOSITION 4.9. $\rho_{N_1,\lambda} = \rho_{N_2,\lambda} \in \mathbb{Q} \Longrightarrow g_{N_1,\lambda}$ and $g_{N_2,\lambda}$ have the same stable I/O pattern.

Proof - Suppose that $g_{N_1,\lambda}$ and $g_{N_2,\lambda}$ have different stable I/O patterns. Because of Proposition 4.8, at least an **a** and an **ac** will have their positions permuted. Without loss of generality, we can assume that $N_1 < N_2$ and that position k is the first one in which both patterns differ:

$$\hat{g}^{k-1}_{N_1,\lambda}(N_1) < q + \lambda \quad , \quad \hat{g}^{k-1}_{N_2,\lambda}(N_2) \geq q + 1, \quad q \in \mathbb{N}$$

Since g_λ is continuous and monotonically increasing with N:

$$\exists N^* \text{ s.t. } N_1 < N^* < N_2 \;\wedge\; q + \lambda \leq \hat{g}^{k-1}_{N^*,\lambda}(N^*) < q + 1$$

The stable I/O pattern of $g_{N^*,\lambda}$ has a **b** where that of $g_{N_1,\lambda}$ has an **a** and that of $g_{N_2,\lambda}$ has an **ac**. Hence, $g_{N^*,\lambda}(\emptyset)=\emptyset$ has a solution of lower period than those of $g_{N_i,\lambda}(\emptyset)=\emptyset$, i=1,2, and consequently:

$$\rho_{N^*,\lambda} \neq \rho_{N_i,\lambda}, \quad i = 1,2.$$

But since ρg_λ increases monotonically with N and $\rho_{N_1,\lambda}=\rho_{N_2,\lambda}$:

$$\rho_{N_1,\lambda} = \rho_{N,\lambda} = \rho_{N_2,\lambda}, \quad \forall N \in [N_1, N_2]$$

Thus, we have come to a contradiction.∎

PROPOSITION 4.10. If there exists stable (s:r) entrainment, the region of the (N,λ)-parameter space where it takes place is:

$$\frac{(r-1+\lambda)\,\lambda^{s-1} + (1-\lambda)\sum_{i=0}^{s-2}\lambda^i\sum_{j=1}^{i}\mu_j}{\sum_{i=0}^{s-1}\lambda^i} < N < \frac{r\,\lambda^{s-1} + (1-\lambda)\sum_{i=1}^{s-2}\lambda^i\sum_{j=1}^{i}\mu_j}{\sum_{i=0}^{s-1}\lambda^i} \qquad (4.12)$$

where μ_j is defined as in equation 4.10.

Proof - Following the same reasoning as in the proof of Proposition 4.9, we arrive at equation 4.11. Since $\langle\emptyset_s\rangle_1\in(\lambda,1)$:

$$r - 1 + \lambda < \hat{\emptyset}_s < r$$

Using equation 4.10:

$$r-1+\lambda < \sum_{i=0}^{s-1}\frac{1}{\lambda^i}N - (1-\lambda)\sum_{i=1}^{s-2}\frac{1}{\lambda^{s-i-1}}\sum_{j=1}^{i}\mu_j < r$$

Isolating N from the last expresion, we obtain the desired inequality. Any N that yields stable (s:r) entrainment will then satisfy that condition. Notice that the condition is unique because of Proposition 4.9, which assures that the values μ_j, j=1,... s, will be constant for each entrainment ratio.

We still have to prove that, if stable (s:r) entrainment takes place at all, any N belonging to the stated interval will yield it. It suffices to show that μ_j, j=1,... s, remains constant when N varies within this interval.

Suppose that N^* gives place to stable (s:r) entrainment and that $N^*+\varepsilon^*$, with $\varepsilon^* \in \mathbb{R}^+$, is the minimum value greater than N^* and belonging to the interval, for which a change in the I/O pattern takes place. Because of the continuity and monotonicity of $\hat{g}_\lambda$, this implies the transformation of one **ac** into a **b** and, in terms of phases, there exists k<s s.t.

$$\sum_{j=1}^{k-2} \mu_j + \lambda(1-(N^*+\varepsilon)) < \hat{g}^{k-1}_{\lambda,N^*+\varepsilon}(N^*+\varepsilon) < \sum_{j=1}^{k-2} \mu_j + \lambda, \quad \forall \varepsilon \in \mathbb{R}^+, \ \varepsilon < \varepsilon^*$$

and
$$\hat{g}^{k-1}_{\lambda,N^*+\varepsilon^*}(N^*+\varepsilon^*) = \sum_{j=1}^{k-2} \mu_j + \lambda$$

Therefore, $N^*+\varepsilon^*$ yields stable $(\sum_{j=1}^{k-1} \mu_j : k)$ entrainment, a fact that contradicts the continuity of ρ_{g_λ}.∎

Likewise, the same result could be proved for the values in the stated interval less than N.

Corollary 4.10.1. If stable (s:r) entrainment exists, the length of the interval of parameter N where it takes place is:

$$\frac{\lambda^{s-1} - \lambda^{s}}{\sum_{i=0}^{s-1} \lambda^i}$$

Notice that this length is independent of r.

Because of its resemblance to the Cantor function, a detailed description of which can be found in Gelbaum and Olmsted (1964), we will say that $f: [a,b) \dashrightarrow \mathbb{R}$ is a **generalized Cantor function** if it satisfies the following conditions:

(a) f is continuous and monotonically increasing.

(b) $f(a)=a$ and $\lim_{x \to b} f(x)=b$.

(c) $f'(x)=0$, $x \in [a,b)-C$, where C denotes a set of measure zero.

PROPOSITION 4.11. The mapping

$$\rho_{g_\lambda} : [0,1) \dashrightarrow \mathbb{R}$$
$$N \longmapsto \rho_{N,\lambda}$$

is a generalized Cantor function.

Proof - We have concluded in the preceding section that ρ_{g_λ} is continuous and monotonically increasing with N. Furthermore, $\rho_{g_\lambda}(0)=0$ and, since $\rho_{g_\lambda}(N^*)=1$, $\forall N^* \in (\lambda,1)$, it is also true that $\lim_{N \to 1} \rho_{g_\lambda}(N)=1$.

From all that, we deduce:

$$\forall \frac{r}{s} \in [0,1) \cap \mathbb{Q}, \quad I_{r/s} = \{ N \in [0,1) : \rho_{g\lambda}(N) = \frac{r}{s} \} \neq \emptyset$$
$$(r,s)=1$$

Because of Proposition 4.10 and Corollary 4.10.1, $I_{r/s}$ is an open interval of length:

$$\frac{\lambda^{s-1} - \lambda^{s}}{\sum_{i=0}^{s-1} \lambda^{i}} > 0$$

and, consequently,

$$\rho'_{g\lambda}(x) = 0, \quad \forall x \in I_{r/s}.$$

For each $s \in \mathbb{N}^+$, the number of values of r such that $(r,s)=1$ and $(r/s)<1$ is given by Euler's totient function $\varphi(s)$ (Vinogradov, 1977). Adding up the lengths of all intervals $I_{r/s}$, we obtain:

$$\sum_{s=1}^{\infty} \varphi(s)\left[\frac{\lambda^{s-1} - \lambda^{s}}{\sum_{i=0}^{s-1} \lambda^{i}}\right] = \sum_{s=1}^{\infty} \varphi(s) \left[\frac{\lambda^{s-1}(1-\lambda)}{\frac{1-\lambda^{s}}{1-\lambda}}\right] = \frac{(1-\lambda)^{2}}{\lambda} \sum_{s=1}^{\infty} \varphi(s) \frac{\lambda^{s}}{1-\lambda^{s}} =$$

$$= \frac{(1-\lambda)^{2}}{\lambda} \sum_{s=1}^{\infty} \varphi(s) \sum_{k=1}^{\infty} \lambda^{ks} = \frac{(1-\lambda)^{2}}{\lambda} \sum_{m=1}^{\infty} \lambda^{m} \sum_{l/m} \varphi(l) =$$

$(m = ks)$

$$= \frac{(1-\lambda)^{2}}{\lambda} \sum_{m=1}^{\infty} m\, \lambda^{m} = \frac{(1-\lambda)^{2}}{\lambda} \frac{\lambda}{(1-\lambda)^{2}} = 1$$

where we have used the result: $\sum_{l/m} \varphi(l)=m$, in which the sum is taken over all divisors l of m (Vinogradov, 1977).

Hence, $\bigcup_{\substack{(r,s)=1\\(r/s)<1}} I_{r/s}$ is a set of measure one and therefore ρ_{g_λ} has zero derivative, except in the set $[0,1) - \bigcup I_{r/s}$, which is of measure zero.∎

In Figure 4.6 the functions ρ_{g_λ} corresponding to three different values of λ are represented with a discretization step of $\Delta N=(1/1000)$. Figure 4.7 shows the three-dimensional representation of the function ρ_g in the (N,λ)-parameter space, with discretization steps $\Delta N=(1/300)$ and $\Delta\lambda=(1/10)$. The recursive properties displayed by this function will be analyzed in the next section.

Corollary 4.11.1. The mapping $\bar{\rho}_{g_\lambda}: [0,\infty) \longrightarrow \mathbb{R}$ defined by:

$$\bar{\rho}_{g\lambda}(N) = \rho_{g\lambda}(\langle N\rangle_1) + N - \langle N\rangle_1$$

is a generalized Cantor function.

Summarizing, the set of stimulation frequencies --either higher or lower than the pacemaker's spontaneous one-- not yielding stable entrainment is of measure zero. Furthermore, each entrainment ratio (s/r) is originated by the set of input frequencies of positive measure given by expression 4.12. The well-known regions of stable subharmonic (s:1) and superharmonic (1:r) entrainment (Segundo and Kohn, 1981) are particular cases of this general expression.

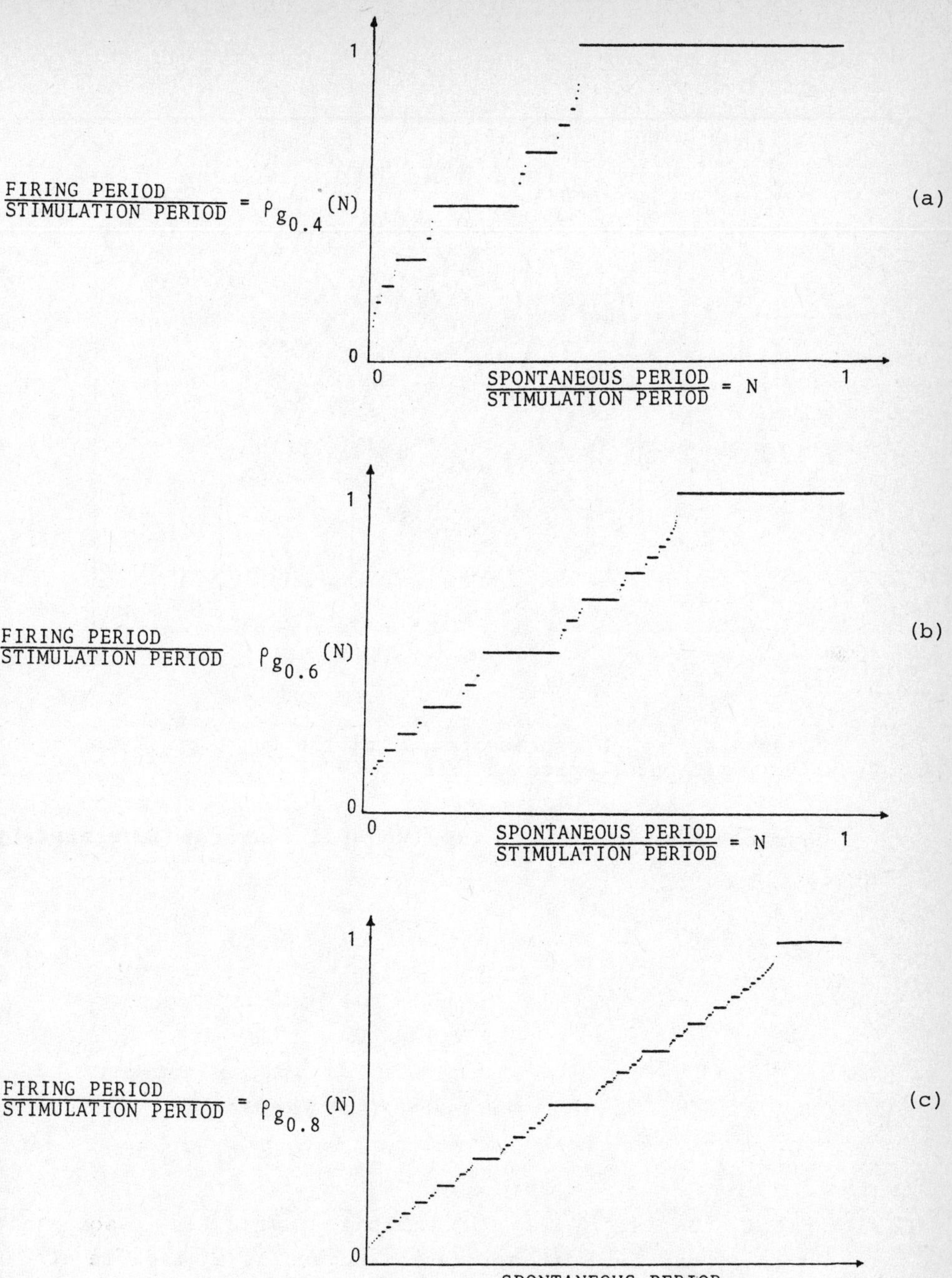

*** Figure 4.6 - Representation of the generalized Cantor function ρ_{g_λ} for the values: (a) $\lambda=0.4$, (b) $\lambda=0.6$ and (c) $\lambda=0.8$.

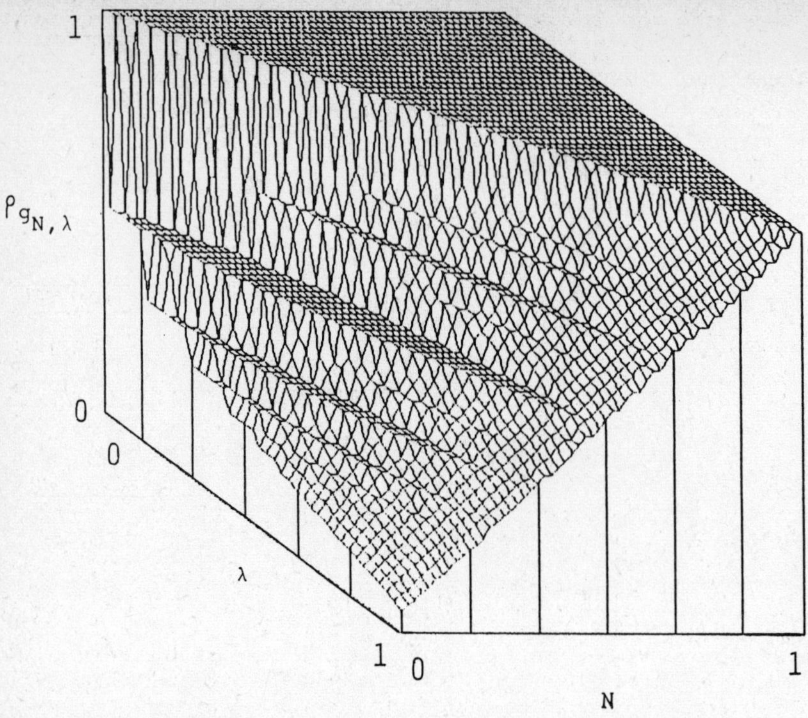

*** Figure 4.7 - 3-D representation of the function ρ_{g_λ} in the (N,λ)-parameter space.

4.3.3.3 - Characterization of the input/output patterns generated by the function $g_{N,\lambda}$.

"No veo con los ojos: las palabras
son mis ojos. Vivimos entre nombres;
lo que no tiene nombre todavía
no existe: Adán de lodo,
no un muñeco de barro, una metáfora.
Ver el mundo es deletrearlo".

O. Paz, 1975

We will stratify the set $[0,1)\cap\mathbb{Q}$ in infinite levels P_i, $i\in\mathbb{N}$, so that the I/O pattern associated with a rotation number of a given level can be deduced from the I/O patterns associated with numbers of lower levels.

The generating rule of the elements of a level from those of lower levels is the same used for the construction of the so-called "reduced fractions" (Vinogradov, 1977) and, for this reason, a set of tools and results from the Elementary Theory of Numbers, like Euclid's algorithm, becomes directly applicable in this context.

Let us define **the first level** $P_1=\left\{\frac{1}{s_1} : s_1 \in \mathbb{N}^+\right\}$. Each rotation number $\frac{1}{s_1} \in P_1$ has an associated I/O pattern $\mathbf{a}^{s_1-1}\mathbf{b}$, which corresponds, as we previously saw, to the stable subharmonic $(s_1:1)$ entrainment. According to Proposition 4.7, its region of existence is:

$$\frac{\lambda^{s_1}}{\sum_{i=0}^{s_1-1} \lambda^i} < N < \frac{\lambda^{s_1-1}}{\sum_{i=0}^{s_1-1} \lambda^i}$$

To define two continuous functions $\rho_{g_\lambda}^{inf_1}$ and $\rho_{g_\lambda}^{sup_1}$: $[0,1) \dashrightarrow [0,1)$, that go through the lower and upper bounds of those intervals of existence, respectively, we must impose:

$$\rho_{g\lambda}^{inf_1}\left(\frac{\lambda^{s_1}}{\sum_{i=0}^{s_1-1} \lambda^i}\right) = \frac{1}{s_1} \qquad \rho_{g\lambda}^{sup_1}\left(\frac{\lambda^{s_1-1}}{\sum_{i=0}^{s_1-1} \lambda^i}\right) = \frac{1}{s_1} \qquad (4.13)$$

and assure the continuity of $\rho_{g_\lambda}^{inf_1}$ and $\rho_{g_\lambda}^{sup_1}$ with respect to the variables x and y, respectively, by defining:

$$x = \frac{\lambda^{s_1}}{\sum_{i=0}^{s_1-1} \lambda^i} = \frac{(1-\lambda)\lambda^{s_1}}{(1-\lambda^{s_1})} ; \quad y = \frac{\lambda^{s_1-1}}{\sum_{i=0}^{s_1-1} \lambda^i} = \frac{(1-\lambda)\ \lambda^{s_1-1}}{(1-\ \lambda^{s_1})}$$

from which:

$$s_1 = \frac{\log(x/1-\lambda+x)}{\log \lambda} \qquad s_1 = \frac{\log(\lambda y/1-\lambda+\lambda y)}{\log \lambda}$$

and, therefore, the expressions 4.13 will end up being:

$$\rho_{g\lambda}^{inf_1}(x) = \frac{\log \lambda}{\log\left(\frac{1}{1+\frac{1-\lambda}{x}}\right)} \qquad \rho_{g\lambda}^{sup_1}(y) = \frac{\log \lambda}{\log\left(\frac{1}{1+\frac{1-\lambda}{\lambda y}}\right)}$$

The graphs of these two functions are shown in Figure 4.8.

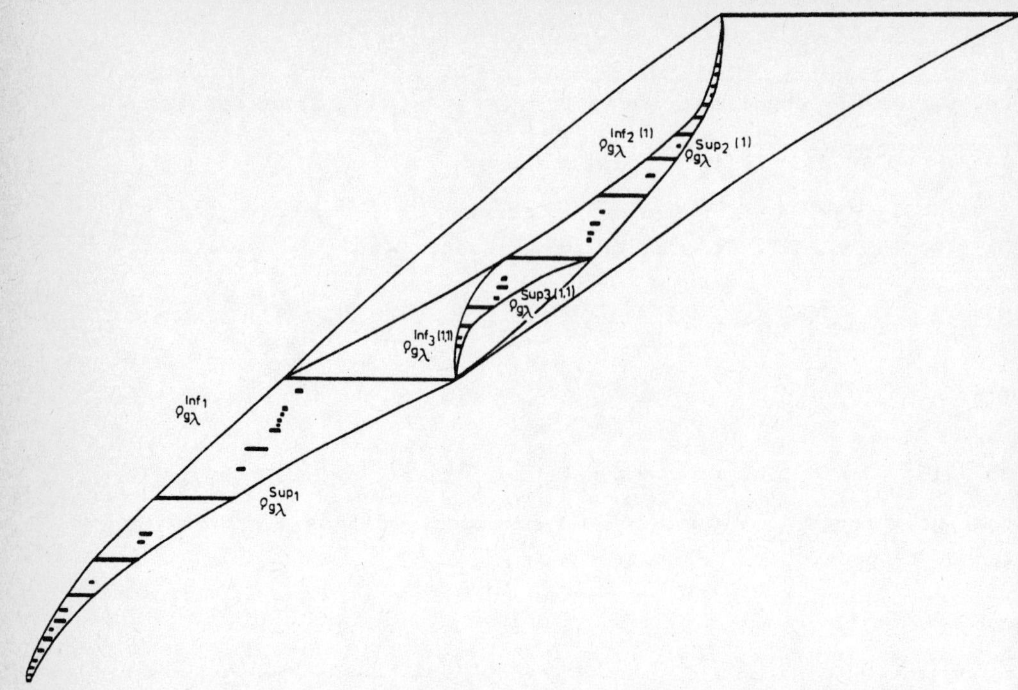

*** Figure 4.8 - Graphical representation of the functions that go through the lower and upper bounds of the entrainment intervals corresponding to the rotation numbers of the first level $(\rho_{g_\lambda}^{inf_1}, \rho_{g_\lambda}^{sup_1})$, second level $(\rho_{g_\lambda}^{inf_2(1)}, \rho_{g_\lambda}^{sup_2(1)})$, and third level $(\rho_{g_\lambda}^{inf_3(1,1)}, \rho_{g_\lambda}^{sup_3(1,1)})$.

We define **the second level** $P_2 = \bigcup_{s_1 \in \mathbb{N}^+} P_2^{s_1}$, with:

$$P_2^{s_1} = \{\frac{s_2}{s_1 s_2 + 1} \; : \; s_2 \in \mathbb{N}^+ \}.$$

Between each two consecutive numbers of the first level, $\frac{1}{s_1+1}$ and $\frac{1}{s_1}$, there are infinite elements of the second level, those belonging to $P_2^{s_1}$, which constitute a strictly increasing sequence with respect to s_2, with limit $\frac{1}{s_1}$. In mathematical terms:

$$\frac{1}{s_1+1} \leqslant \frac{s_2}{s_1 s_2+1} < \frac{s_2+1}{s_1(s_2+1)+1} < \frac{1}{s_1}$$

and (4.14)

$$\lim_{s_2 \to \infty} \frac{s_2}{s_1 s_2 + 1} = \frac{1}{s_1}$$

PROPOSITION 4.12. $\frac{1}{s_1+1} < \rho_{N^*,\lambda} < \frac{1}{s_1} \Longrightarrow$ The partial I/O patterns generated by $g_{N^*,\lambda}$ between consecutive discharges are $\mathbf{a}^{s_1+1}\mathbf{c}$, $\mathbf{a}^{s_1}\mathbf{c}$ and $\mathbf{a}^{s_1}\mathbf{b}$.

Proof - Suppose that $g_{N_1,\lambda}$ and $g_{N_2,\lambda}$ have rotation numbers $\frac{1}{s_1+1}$ and $\frac{1}{s_1}$, respectively. Then,

$$\lambda \leq \hat{g}^{s_1}_{N_1,\lambda}(N_1) < 1$$

$$\lambda \leq \hat{g}^{s_1-1}_{N_2,\lambda}(N_2) < 1$$

From the inequality in the statement of the proposition and the fact that ρ_{g_λ} is a monotonically increasing mapping, we deduce that $N_1 < N^* < N_2$.

Now, since $\hat{g}_\lambda(N,\hat{\emptyset})$ increases monotonically with N and $\hat{\emptyset}$:

$$1 \leq g^{s_1}_{N^*,\lambda}(N^*)$$

$$g^{s_1-1}_{N^*,\lambda}(N^*)$$

Thus,

$$\lambda(1-N^*) \leq \hat{g}^{s_1-1}_{N^*,\lambda}(N^*)$$

$$\hat{g}^{s_1}_{N^*,\lambda}(N^*) < 1 + N^*$$

But $\forall \emptyset < N^*$:

$$\lambda(1-N^*) \leqslant \hat{g}^{s_1-1}_{N^*,\lambda}(N^*) < \hat{g}^{s_1}_{N^*,\lambda}(\emptyset) < \hat{g}^{s_1}_{N^*,\lambda}(N^*) < 1+N^*$$

Taking into account that N^* is the maximum phase at which a stimulation impulse can arrive after discharge of the pacemaker and applying equation 4.9, we obtain the desired result.∎

PROPOSITION 4.13. Each rotation number $\frac{s_2}{s_1 s_2+1}$ has as associated I/O pattern $(\mathbf{a}^{s_1}\mathbf{c})^{s_2-1}\mathbf{a}^{s_1}\mathbf{b}$.

Proof - For $s_2=1$, the result is obvious. If $s_2>1$, inequality 4.14 assures that Proposition 4.12 can be applied and, as a consequence, the final part of the I/O pattern has to be $\mathbf{a}^{s_1}\mathbf{b}$.

Subtracting 1 from the number of discharges (numerator of the rotation number) and s_1+1 from the number of input impulses (denominator of the rotation number), it follows that the pacemaker has to fire (s_2-1) times for every $s_1(s_2-1)$ stimulation impulses. Thus, the only possible partial pattern is $\mathbf{a}^{s_1}\mathbf{c}$, which implies the desired result.∎

According to Proposition 4.10, the existence region of the I/O patterns of this type is:

$$\frac{(r-1+\lambda)\,\lambda^{s_1 s_2} + (1-\lambda)\sum_{i=1}^{s_1 s_2-1}\lambda^i \sum_{j=1}^{i}\mu_j}{\sum_{i=0}^{s_1 s_2}\lambda^i} < N < \frac{r\,\lambda^{s_1 s_2} + (1-\lambda)\sum_{i=1}^{s_1 s_2-1}\lambda^i \sum_{j=1}^{i}\mu_j}{\sum_{i=0}^{s_1 s_2}\lambda^i}$$

$$\text{where } \mu_j = \begin{cases} 1, & \text{if } j=ks_1 \text{ with } k\in\mathbb{N},\ k<s_2 \\ 0, & \text{otherwise} \end{cases}$$

Incorporating the values of μ_j into the inequality:

$$\frac{(r-1+\lambda)\,\lambda^{s_1 s_2} + (1-\lambda) \sum_{k=1}^{s_2-1} \sum_{i=ks_1}^{(k+1)s_1-1} k\,\lambda^i}{\sum_{i=0}^{s_1 s_2} \lambda^i} < N <$$

$$< \frac{r\,\lambda^{s_1 s_2} + (1-\lambda) \sum_{k=1}^{s_2-1} \sum_{i=ks_1}^{(k+1)s_1-1} k\,\lambda^i}{\sum_{i=0}^{s_1 s_2} \lambda^i}$$

$\forall s_1 \in \mathbb{N}^+$, it is possible to define two continuous functions $\rho_{g_\lambda}^{\inf_2(s_1)}$ and $\rho_{g_\lambda}^{\sup_2(s_1)}$: $[0,1) \dashrightarrow [0,1)$, that go through the lower and upper bounds of the intervals of stable $(s_1 s_2+1:s_2)$ entrainment, respectively, following the same procedure used to define $\rho_{g_\lambda}^{\inf_1}$ and $\rho_{g_\lambda}^{\sup_1}$.

Since the analytic expression of these functions is rather complex and has no particular theoretical or practical interest, we will limit ourselves to showing their graphs, for $s_1=1$, in Figure 4.8.

Next, we define **the third level** $P_3 = \bigcup_{s_1 \in \mathbb{N}^+} \bigcup_{s_2 \in \mathbb{N}^+} P_3^{s_1,s_2}$, with:

$$P_3^{s_1,s_2} = \left\{ \frac{s_2 s_3 - 1}{(s_1 s_2+1)s_3 + s_1} : s_3 \in \mathbb{N}^+ \right\}.$$

Between each two consecutive numbers of the second level $\frac{s_2}{s_1 s_2\ 1}$ and $\frac{s_2+1}{s_1(s_2+1)+1}$, there are infinite elements of the third level, those belonging to $P_3^{s_1,s_2}$, which form a strictly increasing sequence with respect to s_3, with limit $\frac{s_2}{s_1 s_2+1}$. In mathematical terms:

$$\frac{s_2}{s_1 s_2+1} < \frac{s_2(s_3+1)+1}{(s_1 s_2+1)(s_3+1)+s_1} < \frac{s_2 s_3+1}{(s_1 s_2+1)s_3+s_1} \leqslant \frac{s_2+1}{s_1(s_2+1)+1}$$

and (4.15)

$$\lim_{s_3\to\infty} \frac{s_2s_3+1}{(s_1s_2+1)s_3+s_1} = \frac{s_2}{s_1s_2+1}$$

PROPOSITION 4.14. Each rotation number $\frac{s_2s_3+1}{(s_1s_2+1)s_3+s_1} \in P_3$ has as associated I/O pattern $((\mathbf{a}^{s_1}\mathbf{c})^{s_2}\mathbf{a})^{s_3-1}(\mathbf{a}^{s_1}\mathbf{c})^{s_2}\mathbf{a}^{s_1}\mathbf{b}$.

Proof - For $s_3=1$, the result is obvious. Now suppose that:

$$\rho_{N^*,\lambda}= \frac{s_2s_3+1}{(s_1s_2+1)s_3+s_1}, \quad s_3 \in \mathbb{N}^+, \quad s_3 > 1$$

From inequality 4.15 and the increasing character of ρg_λ, $\hat{g}_\lambda$ and $\hat{g}_{N^*,\lambda}$, it follows that:

$$s_2 \leqslant \hat{g}^{s_1s_2}_{N^*,\lambda}(N^*)$$

$$\hat{g}^{ks_1s_2+s_1+(k-1)}_{N^*,\lambda}(N^*) < \lambda+ ks_2, \quad k \in \mathbb{N}^+, \quad k < s_3 \tag{4.16}$$

According to inequalities 4.14 and 4.15, $\frac{1}{s_1+1} < \rho_{N^*,\lambda} < \frac{1}{s_1}$, and because of Proposition 4.12, the minimum and maximum number of input impulses between consecutive discharges are s_1 and s_1+1, therefore the initial part of the I/O pattern will be $(\mathbf{a}^{s_1}\mathbf{c})^{s_2}\mathbf{a}$ and it will hold that:

$$s_2 + N^* \leqslant g^{s_1s_2+1}_{N^*,\lambda}(N^*)$$

Since $g_{N^*,\lambda}$ is an increasing function, the minimum number of blocks $\mathbf{a}^{s_1}\mathbf{c}$ that will follow the above partial pattern will be s_2, which is the maximum number as well because of inequality 4.16 for k=2:

$$\hat{g}^{2s_1s_2+s_1+1}_{N^*,\lambda}(N^*) < \lambda +2s_2$$

Thus,

$$2s_2 + N^* \leqslant g^{2s_1s_2+2}_{N^*,\lambda}(N^*)$$

And by induction:

$$ks_2 + N^* \leqslant \hat{g}^{ks_1 s_2 + k}_{N^*,\lambda}(N^*) \quad , \forall k \in \mathbb{N}^+ \ , \ k < s_3 \qquad (4.17)$$

Up to $k=s_3-1$, the I/O pattern will be $((\mathbf{a}^{s_1}\mathbf{c})^{s_2}\mathbf{a})^{s_3-1}$. But, from Proposition B.4:

$$s_2 s_3 + \lambda \leqslant \hat{g}^{s_3 s_1 s_2 + (s_3-1) + s_1}_{N^*,\lambda}(N^*) < s_2 s_3 + 1$$

from which, and together with inequality 4.17 for $k=s_3-1$, the final part of the pattern is derived: $(\mathbf{a}^{s_1}\mathbf{c})^{s_2}\mathbf{a}^{s}\ \mathbf{b}.$∎

Figure 4.8 shows the graphical representation of the functions $\rho_{g\lambda}^{inf_3(s_1,s_2)}$ and $\rho_{g\lambda}^{sup_3(s_1,s_2)}$, defined in an analogous way to those corresponding to the first and second level, for $s_1=s_2=1$.

Finally, we define **the nth level** $P_n = \bigcup_{(s_1,\ldots,s_{n-1}) \in (\mathbb{N}^+)^{n-1}} P_n^{s_1,\ldots,s_{n-1}}$, with:

$$P_n^{s_1,\ldots,s_{n-1}} = \{p^{s_1,\ldots,s_{n-1}}(s_n) = \frac{A_{n-1}\,s_n + A_{n-2}}{B_{n-1}\,s_n + B_{n-2}} : s_n \in \mathbb{N}^+\} \qquad (4.18)$$

$$\frac{A_{n-1}}{B_{n-1}} \in P_{n-1}^{s_1,\ldots,s_{n-2}} \quad \wedge \quad \frac{A_{n-2}}{B_{n-2}} \in P_{n-2}^{s_1,\ldots,s_{n-3}}$$

As anticipated at the beginning of this section, the recursive generating rule expressed by equation 4.18 is the same one used for the construction of the reduced fractions (Vinogradov, 1977). In the present case, the fractions will belong to the interval [0,1], since we have implicitely taken as seeds the values $A_0=0$, $B_0=1$, $A_1=1$ and $B_1=s_1$. If we had taken instead $A_0=1$, $B_0=0$, $A_1=s_1$ and $B_1=1$, the generated values would have been all the rational numbers. We thus have:

$$\frac{A_0}{B_0} = \frac{0}{1}$$

$$\frac{A_1}{B_1} = \frac{1}{s_1} \quad , \quad \forall s_1 \in \mathbb{N}^+$$

$$\frac{A_2}{B_2} = \frac{A_1 s_2 + A_0}{B_1 s_2 + B_0} = \frac{s_2}{s_1 s_2 + 1} = \frac{1}{s_1 + \frac{1}{s_2}} \quad , \quad \forall s_2 \in \mathbb{N}^+$$

$$\frac{A_3}{B_3} = \frac{A_2 s_3 + A_1}{B_2 s_3 + B_1} = \frac{s_2 s_3 + 1}{(s_1 s_2 + 1) s_3 + s_1} = \frac{1}{s_1 + \frac{1}{s_2 + \frac{1}{s_3}}} \quad , \quad \forall s_3 \in \mathbb{N}^+$$

and, in the generic case,

$$\frac{A_n}{B_n} = \frac{A_{n-1} s_n + A_{n-2}}{B_{n-1} s_n + B_{n-2}} = \cfrac{1}{s_1 + \cfrac{1}{s_2 + \cfrac{1}{s_3 + \ddots + \cfrac{1}{s_n}}}} \quad , \quad \forall s_n \in \mathbb{N}^+$$

This identification has some useful consequences:

(a) All fractions $\frac{A_n}{B_n}$ are irreducible.

(b) Between every two consecutive numbers of level n-1:

$$p^{s_1, \ldots, s_{n-2}}(s_{n-1}),\ p^{s_1, \ldots, s_{n-2}}(s_{n-1}+1) \in P_{n-1}^{s_1, \ldots, s_{n-2}}$$

there are infinite numbers of level n:

$$p^{s_1, \ldots, s_{n-1}}(s_n) \in P_n^{s_1, \ldots, s_{n-1}}$$

which constitute a strictly increasing or decreasing sequence with respect to s_n, depending on whether n is even or odd, with limit $p^{s_1, \ldots, s_{n-2}}(s_{n-1})$. In mathematical terms:

if n is even,

$$p^{s_1, \ldots, s_{n-2}}(s_{n-1}+1) \leqslant p^{s_1, \ldots, s_{n-1}}(s_n) < \quad (4.19)$$

$$< p^{s_1, \ldots, s_{n-1}}(s_n+1) < p^{s_1, \ldots, s_{n-2}}(s_{n-1})$$

if, on the contrary, n is odd,

$$p^{s_1,\dots,s_{n-2}}(s_{n-1}) < p^{s_1,\dots,s_{n-1}}(s_n+1) < \tag{4.20}$$

$$< p^{s_1,\dots,s_{n-1}}(s_n) \leqslant p^{s_1,\dots,s_{n-2}}(s_{n-1}+1)$$

and in both cases:

$$\lim_{s_n\to\infty} p^{s_1,\dots,s_{n-1}}(s_n) = p^{s_1,\dots,s_{n-2}}(s_{n-1})$$

(c) $\bigcup_{n=1}^{\infty} P_n = \mathbb{Q} \cap [0,1]$ (4.21)

(d) $\lim_{n\to\infty} \left[p^{s_1,\dots,s_{n-1}}(s_n+1) - p^{s_1,\dots,s_{n-1}}(s_n) \right] = 0,$

$\forall s_i \in \mathbb{N}^+ , \quad \forall i \in \mathbb{N}^+$

(e) Euclid's algorithm allows one to find the n-tuple $(s_1,\dots\ s_n)$ which characterizes the position occupied by each irreducible fraction $(r/s) \in [0,1]$ in the stratification on levels that we have just described:

$$s = s_1 r + r_1 \quad , \quad r_1 < r$$

$$r = s_2 r_1 + r_2 \quad , \quad r_2 < r_1$$

$$r_1 = s_3 r_2 + r_3 \quad , \quad r_3 < r_2$$

$$\vdots$$

$$r_{n-2} = s_n r_{n-1} + 0 , \ 0 < r_{n-1}$$

Next, we will prove that the aforementioned n-tuple uniquely determines the I/O pattern associated with each rotation number:

$$\rho_{N,\lambda} = r/s = p^{s_1,\dots,s_{n-1}}(s_n) .$$

Given an I/O pattern $\eta = \eta_1 \mathbf{b}$, associated with a rotation number of level n, we define its corresponding intermediate pattern α_η in the following way:

$$\alpha_\eta = \begin{cases} \eta_1 \mathbf{ac}, \text{ if n is even} \\ \eta_1 \mathbf{ca}, \text{ if n is odd} \end{cases}$$

PROPOSITION 4.15. If η and ω are I/O patterns associated with rotation numbers $p^{s_1,\ldots,s_{n-2}}(s_{n-1}) \in P_{n-1}^{s_1,\ldots,s_{n-2}}$ and $p^{s_1,\ldots,s_{n-1}}(s_n-1) \in$ $\in P_n^{s_1,\ldots,s_{n-1}}$, respectively, then the I/O pattern associated with $p^{s_1,\ldots,s_{n-1}}(s_n) \in P_n^{s_1,\ldots,s_{n-1}}$ will be $\alpha_\eta \omega$.

Proof - Let's suppose that the mapping $g_{N^*,\lambda}$ has rotation number:

$$p^{s_1,\ldots,s_{n-1}}(s_n) = \frac{A_{n-1}\, s_n + A_{n-2}}{B_{n-1}\, s_n + B_{n-2}}$$

If n is even, from inequality 4.19 and the continuity and monotonicity of ρg_λ and $\hat{g}_\lambda$, we have:

$$\hat{g}_{N^*,\lambda}^{B_{n-1}-1}(N^*) < \lambda + A_{n-1} - 1$$

Because of equation 4.18, there is no intermediate pattern of period lower than B_n between those corresponding to $p^{s_1,\ldots s_{n-2}}(s_{n-1})$ and $p^{s_1,\ldots,s_{n-1}}(s_n)$, and therefore the initial part of the I/O pattern will be α_η.

Likewise, if n is odd, from inequality 4.20 we obtain:

$$A_{n-1} \leqslant \hat{g}_{N^*,\lambda}^{B_{n-1}-1}(N^*)$$

and, by the same reasoning used before, the initial part of the I/O pattern associated with $p^{s_1,\ldots,s_{n-1}}(s_n)$ will be α_η.

α_η contains A_{n-1} pacemaker discharges and B_{n-1} stimulation impulses. Subtracting these quantities from the numerator and the denominator of $p^{s_1,\ldots s_{n-1}}(s_n)$, we obtain $p^{s_1,\ldots,s_{n-1}}(s_n-1)$.

It remains to be proved that the only possible final pattern is precisely the one associated with $p^{s_1,\dots s_{n-1}}(s_n-1)$. From Propositions 4.10 and 4.11:

$$\forall \emptyset_f \in [\lambda,1) \;\wedge\; \forall \underset{\substack{(r,s)=1\\ s>0}}{\frac{r}{s} \in \mathbb{Q}}\;,\; \exists N' \in [0,1) \text{ s.t. } \hat{g}^{s-1}_{N',\lambda}(N')=\emptyset_f+r-1$$

In particular, $\forall \emptyset_f \in [\lambda,1)$, $\exists N_1', N_2' \in [0,1)$ s.t.

$$\hat{g}^{B_{n-1}(s_n-1)+B_{n-2}-1}_{N_1',\lambda}(N_1') = \emptyset_f + A_{n-1}(s_n-1) + A_{n-2} - 1$$

$$\hat{g}^{B_{n-1}s_n+B_{n-2}-1}_{N_2',\lambda}(N_2') = \emptyset_f + A_{n-1}s_n + A_{n-2} - 1 \tag{4.22}$$

Let's assume that the final part of the I/O pattern associated with $p^{s_1,\dots,s_{n-1}}(s_n)$ is different from the complete I/O pattern associated with $p^{s_1,\dots,s_{n-1}}(s_n-1)$, and that $k\in\mathbb{N}^+$ $(k<B_{n-1}(s_n-1)+B_{n-2}-1)$ is the first position from the end in which both patterns differ. The difference must necessarily consist of a permutation between an **a** and an **ac**.

If n is even, $N_1'<N_2'$ and, consequently, since $\hat{g}_\lambda$ is monotonically increasing with N:

$$\hat{g}^{B_{n-1}(s_n-1)+B_{n-2}-k}_{N_1',\lambda}(N_1') < A_{n-1}(s_n-1) + A_{n-2} - l + 1$$

$$, l \in \mathbb{N}^+ \tag{4.23}$$

$$\hat{g}^{B_{n-1}s_n+B_{n-2}-k}_{N_2',\lambda}(N_2') < A_{n-1}s_n + A_{n-2} - l + \lambda$$

From the continuity of $\hat{g}_\lambda$ and equations 4.22 and 4.23:

$$\exists N'^* \in [0,1) \text{ s.t. } \hat{g}^{k-1}_{N'^*,\lambda}(N'^*) = \emptyset_f + l - 1$$

Therefore, there exists an I/O pattern, between those corresponding to $p^{s_1,\dots,s_{n-1}}(s_n-1)$ and $p^{s_1,\dots,s_{n-1}}(s_n)$, of lower period than them. This fact contradicts equation 4.18.

If n is odd, $N_1' > N_2'$ and

$$\hat{g}_{N_1',\lambda}^{B_{n-1}(s_n-1)+B_{n-2}-k}(N_1') < A_{n-1}(s_n-1) + A_{n-2} - l + \lambda$$

$$, l \in \mathbb{N}^+$$

$$\hat{g}_{N_2',\lambda}^{B_{n-1}s_n + B_{n-2}-k}(N_2') > A_{n-1}s_n + A_{n-2} - l + 1$$

The previous reasoning also applies to this situation.▮

Corollary 4.15.1. The I/O pattern ζ, associated with a rotation number of level n: $p^{s_1,\dots s_{n-1}}(s_n)$, can be obtained from the I/O patterns η and ω, associated with rotation numbers of level n-1: $p^{s_1,\dots,s_{n-2}}(s_{n-1})$ and $p^{s_1,\dots s_{n-2}}(s_{n-1}+1)$, in the following way:

$$\zeta = \alpha_{\eta}^{s_n-1}\omega$$

Proof - The stated result derives from Proposition 4.15 and the fact that, because of expression 4.18,

$$p^{s_1,\dots,s_{n-1}}(1) = p^{s_1,\dots,s_{n-2}}(s_{n-1}+1)\,.▮$$

The recursive rule expressed in the above Corollary, together with Euclid's algorithm, configure an effective procedure to compute the I/O pattern associated with any rotation number p. As we have previously seen, the aforementioned algorithm allows one to find the n-tuple $(s_1,\dots,s_n)$ that characterizes p and, from it, the following set of generating rules produces the desired I/O pattern, which we will denote $\eta(s_1,\dots,s_n)$:

$$\eta(s_1,\dots s_n) \rightarrow \eta_1(s_1,\dots s_n)\,\mathbf{b}$$

$$\eta_1(s_1,\dots s_{2k}) \rightarrow (\eta_1(s_1,\dots s_{2k-1})\,\mathbf{ac})^{s_{2k}-1}\,\eta_1(s_1,\dots(s_{2k-1}+1))$$

$$\eta_1(s_1,..s_{2k+1}) \rightarrow (\eta_1(s_1,... s_{2k})\mathbf{ca})^{s_{2k+1}-1}\ \eta_1(s_1,...(s_{2k}+1))$$

$$\eta(s_1) \longrightarrow \mathbf{a}^{s_1-1}$$

The availability of an effective procedure to compute the stable I/O pattern for each entrainment ratio assures that the values of the μ_j's which determine the boundaries of the entrainment regions (Proposition 4.10) can actually be derived. From the expression of these boundaries, it is possible to define, for each set $P_n^{s_1,... s_{n-1}}$ of the established stratification (equation 4.18), a pair of continuous functions: one that goes through all the inferior boundaries corresponding to entrainment ratios which belong to that set and another that goes through all the respective superior boundaries. Three instances of these pairs of continuous functions, for sets of levels one, two and three, have been shown in Figure 4.8. In addition to clarifying the form taken by the established stratification, the figure highlights the recursive structure of the generalized Cantor function obtained in the preceding section.

Example - We will generate the I/O pattern associated with the rotation number 23/41. Applying Euclid's algorithm:

$$\begin{aligned} 41 &= 1.23 + 18 \\ 23 &= 1.18 + 5 \\ 18 &= 3.5 \ \ + 3 \\ 5 &= 1.3 \ \ + 2 \\ 3 &= 1.2 \ \ + 1 \\ 2 &= 2.1 \ \ + 0 \end{aligned}$$

The number 23/41 belongs to level 6 and is uniquely determined by the 6-tuple (1,1,3,1,1,2).

$$\begin{matrix} \frac{1}{2} \leqslant & \frac{1}{2} < & \frac{5}{9} \leqslant & \frac{5}{9} < & \frac{14}{25} \leqslant & \frac{23}{41} < & \frac{9}{16} \leqslant & \frac{9}{16} < & \frac{4}{7} \leqslant & \frac{2}{3} < & \frac{1}{1} \\ \in & \in & \in & \in & \in & \in & \in & \in & \in & \in & \in \\ P_1 & P_2 & P_3 & P_4 & P_5 & P_6 & P_5 & P_4 & P_3 & P_2 & P_1 \end{matrix}$$

The chain of generating rules that gives rise to its associated I/O pattern is:

$$\eta(1,1,3,1,1,2) \longrightarrow$$

$$\longrightarrow \eta_1\ (1,1,3,1,1,2)\mathrm{b} \longrightarrow (\eta_1(1,1,3,1,1)\mathrm{ac})^1\ \ \eta_1(1,1,3,1,2)\mathrm{b}$$

$$\longrightarrow ((\eta_1(1,1,3,1)ca)^0 \;\; \eta_1(1,1,3,2)ac)^1 \;\; (\eta_1(1,1,3,1)ca)^1 \;\; \eta_1(1,1,3,2)b$$

$$\longrightarrow ((\eta_1(1,1,3)ac)^1 \;\; \eta_1(1,1,4)ac)^1 \;\; ((\eta_1(1,1,3)ac)^0 \;\; \eta_1(1,1,4)ca)^1$$

$$(\eta_1(1,1,3)ac)^1 \;\; \eta_1(1,1,4)b$$

$$\longrightarrow (\eta_1(1,1)ca)^2 \;\; \eta_1(1,2)ac \;\; (\eta_1(1,1)ca)^3 \;\; \eta_1(1,2)ac \;\; (\eta_1(1,1)ca)^3 \;\; \eta_1(1,2)ca$$

$$(\eta_1(1,1)ca)^2 \;\; \eta_1(1,2)ac \;\; (\eta_1(1,1)ca)^3 \;\; \eta_1(1,2)b$$

$$\longrightarrow ((\eta_1(1)ac)^0 \;\; \eta_1(2)ca)^2 \;\; (\eta_1(1)ac)^1 \;\; \eta_1(2)ac \;\; ((\eta_1(1)ac)^0 \;\; \eta_1(2)ca)^3 \;\; (\eta_1(1)ac)^1$$

$$\eta_1(2)ac \;\; ((\eta_1(1)ac)^0 \;\; \eta_1(2)ca)^3 \;\; (\eta_1(1)ac)^1 \;\; \eta_1(2)ca \;\; ((\eta_1(1)ac)^0 \;\; \eta_1(2)ca)^2$$

$$(\eta_1(1)ac)^1 \;\; \eta_1(2)ac \;\; ((\eta_1(1)ac)^0 \;\; \eta_1(2)ca)^3 \;\; (\eta_1(1)ac)^1 \;\; \eta_1(2)b$$

$$\longrightarrow (aca)^2(ac) \;\; aac(aca)^3acaac(aca)^3acaca(aca)^2ac \;\; aac(aca)^3acab =$$

$$= acaacaacaacacaacaacaacacaacaacaacacaacaacaacaacacaacaacaacab$$

"... Ver el mundo es deletrearlo".

O. Paz, 1975

4.4 - EFFECT OF RANDOMNESS AND LEARNING UPON ENTRAINMENT. FUTURE PROSPECTS.

"El estudio de los llamados azares va ampliando las bandas del billar". *J. Cortázar, 1967*

The analytic study carried out in the present chapter presupposes two ideal deterministic oscillators: the perturbing one and the perturbed one. In modelling, randomness can be introduced in either one of these oscillators or in both simultaneously. If the source of randomness is the perturbing oscillator, the study could be generalized adding a noise term to the $\varnothing_i$ in equation 4.3. If, on the contrary, the source of randomness is the perturbed oscillator, it is necessary to distinguish between two circumstances. When the PRC is invariant to fluctuations in the interspike interval, the analysis can be reduced to the previous case. When the PRC varies as a function of the interspike interval, the emerging situation defies an analytic treatment such as the one we have given, since the phase transition equation changes in each interval.

This last circumstance occurs for the model proposed in the preceding chapter, because the parameter λ depends on the random variable Pb_1 (Figure 4.9). We will let the systematic study of the modifications introduced in the regions and patterns of entrainment when σ_{Pb_1} increases be a subject for future research, limiting ourselves for now to sketching the trends observed in the simulations carried out.

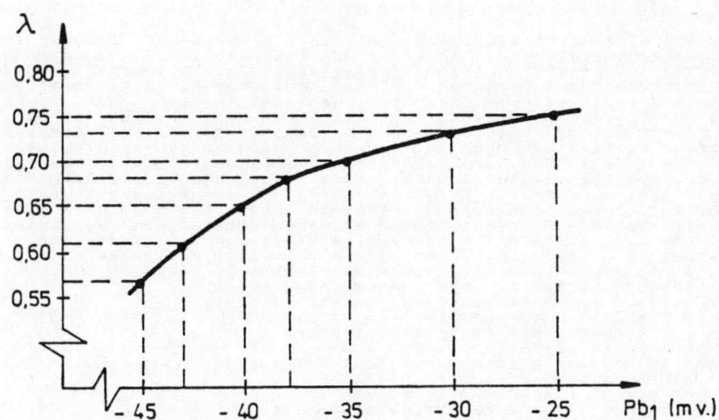

*** Figure 4.9 - Values taken by the parameter λ (phase at which the maximum phase-advancement takes place) that characterizes the PRC, as a function of the asymptotic limit of the spontaneous potential Pb_1.

We first note that the I/O patterns obtained with the deterministic version of the model --explored for discretization time steps of up to 1 ms-- are identical to those deduced analytically for the simplified model. This agreement gives support to the conjecture, based on the topologic character of the properties and arguments used throughout the proofs contained in the present chapter, that the results derived for a piecewise linear PRC can be generalized to a PRC less restrained in the first segment, in which it is only required to be strictly increasing and convex.

The type of PRC considered in the present work falls somewhere in between those studied by Herman (1977) and by Keener (1980). Herman deals with PRCs continuous at least up to the second derivative, while Keener, in the opposite extreme, deals with PRCs that have themselves a discontinuity. Our PRCs are continuous, but have a discontinuity in their first derivative. In the "Conclusions" section of this chapter, we will point out the differences between our results and those found by the two above-mentioned authors.

Figure 4.10 shows the I/O patterns obtained when submitting the stochastic model, for three different values of σ_{Pb_1}, to stimuli of periods between 5 ms and the spontaneous one, for a discretization step of 5 ms. Since the duration of all the strings is the same (5 s), their length, counting only the **a**'s and **b**'s, becomes an indicator of the stimulation frequency and, counting only the **b**'s and **c**'s, becomes an indicator of the firing frequency.

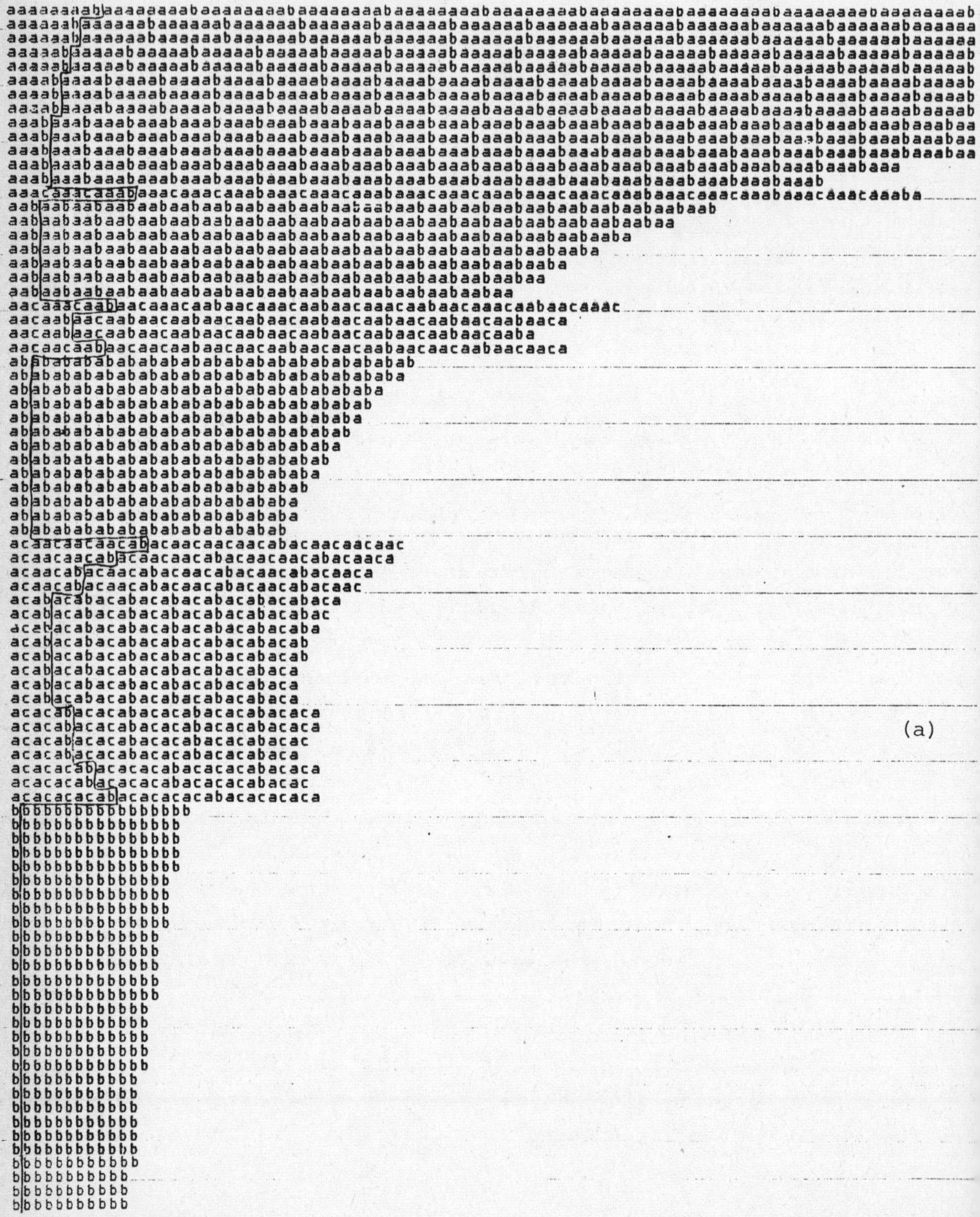

(a)

```
aaaaaaabaaaaaaabaaaaaaabaaaaaaabaaaaaaabaaaaaaabaaaaaaabaaaaaaabaaaaaaabaaaaaaabaaaaaaab
aaaaaabaaaaaabaaaaaabaaaaaabaaaaaabaaaaaabaaaaaabaaaaaabaaaaaabaaaaaabaaaaaabaaaaaabaaaaaa
aaaaaabaaaaaabaaaaabaaaaaabaaaaabaaaaaabaaaaaabaaaaabaaaaaabaaaaaabaaaaabaaaaaabaaaaabaaaa
aaaaabaaaaabaaaaabaaaaabaaaaabaaaaabaaaaabaaaaabaaaaabaaaaabaaaaabaaaaabaaaaabaaaaabaaaaab
aaaaabaaaaabaaaaabaaaaabaaaaabaaaaabaaaaabaaaaabaaaaabaaaaabaaaaacaaaaabaaaabaaaaabaaaaabaa
aacabaaaabaaaabaaaabaaaabaaaabaaaabaaaabaaaabaaaabaaaabaaaabaaaabaaaabaaaabaaaabaaaabaaaab
aaaabaaaabaaaabaaaabaaaabaaaabaaaabaaaabaaaabaaaabaaaabaaaabaaaabaaaabaaaabaaaabaaaabaaaab
aaaabaaaabaaaabaaaacaaaabaaaabaaaabaaaabaaaacaaaabaaaabaaaabaaaacaaaabaaaabaaaabaaaacaaaab
aaabaaabaaabaaabaaabaaabaaabaaaabaaabaaabaaabaaabaaabaaabaaabaaabaaaabaaabaaabaaabaaab
aaabaaabaaabaaabaaabaaabaaabaaabaaabaaabaaabaaabaaabaaabaaabaaabaaabaaabaaabaaabaaabaa
aaabaaabaaabaaabaaabaaabaaabaaabaaabaaabaaabaaabaaabaaabaaabaaabaaabaaabaaabaaabaaabaa
aaabaaabaaabaaabaaabaaabaaabaaabaaabaaabaaabaaabaaabaaabaaabaaabaaabaaabaaabaaa
aaabaaabaaabaaabaaabaaacaaabaaabaaabaaabaaacaaabaaacaaabaaacaaabaaabaaacaaabaaaba
aaacaaabaabaabaaacaaabaaacaaabaaacaaabaaacaaabaaacaaabaabaabaabaaabaabaabaaaca
aabaabaabaabaabaabaabaabaabaabaabaabaabaabaabaabaabaabaabaabaaabaa
aabaabaabaabaabaabaabaabaabaabaabaabaabaabaabaabaabaabaabaabaa
aabaabaabaabaabaabaabaabaabaabaabaabaabaabaabaabaabaabaaba
aabaabaabaabaabaabaabaabaabaabaabaabaabaabaabaabaabaaba
aacaaabaabaabaabaabaabaabaabaabaabaabaabaabaabaaba
aabaabaabaacaaabaabaabaabaabaabaacaaabaabaabaabaabaa
aabaabaabaacaabaacaabaabaabaabaabaabaacaabaabaacaab
aacaabaabaabaabaabaabaabaacaabaacaabaacaabaacaabaa
aacaacaabaacaabaacaabaabaacaaacaacaaacaabaacaabaacaab
abaacaabaacaabaacaabaacaabaacaabaacaabaacaabaacaa
ababaacaabaacaacaabaacaacaababaacaabaacaacaababa
aababaacaabaacababaacaabababababababababa
abababababababababababababaacaaca
abababababababababababababababa
ababababababababababababababab
abababababababababababababa
ababababababababababababab
abababababababababababa
abababababababababab
abababababababababa
ababababababababab
acaababababababababacaacaacaa
ababacaacabacaacaacabababababa
abacaacaacabababababacaacaacab
abacaabacaacabacaacabacaacaacabab
abababacaacabacaacabacaacabacaaca
acaacaacacaacababacaacabacaacabaca
acaacabacababacaacaacabacaacabac
acabacabacaacabacabacaacabacaaca
acabacabacabacaacababacabacab
acabacaacabacabacabacabacabab
acababacabacabacabacabacabac
acacaacabacabacacaacacaacacabaca
acabacabacabacabacabacabaca
acacabacacaacabacabacabacacaac
acacaacabacacabacacabacabacaca
acabacacabacacabacabacacabac
acacabacacabacabacacabacabac
acacabacacabacacabacacabacac
acacacabacacacabacacabacacab
acacacaacacacabacacacabacacaca
bacacabacacacabacacacabacac
bbacacabbacacacabbacacaca
acacacacabbbacacacacabbbb
bacacabbbbacbacbacbbac
bacbbacacacacacacbbacacaca
bbbbbbacacacacabbbbb
bbbbbbbbbbbbbbb
bbbbbbbbacbbbacb
bbbbbbbbbbbbbb
bbbbbbbbbbbbbb
bbbbbbbbbbbbbb
bbbbbbbbbbbbb
bbbbbbbbbbbbb
bbbbbbbbbbbbb
bbbbbbbbbbbbb
bbbbbbbbbbbbb
bbbbbbbbbbbb
bbbbbbbbbbbb
bbbbbbbbbbbb
bbbbbbbbbbbb
bbbbbbbbbbbb
bbbbbbbbbbb
bbbbbbbbbbb
bbbbbbbbbbb
bbbbbbbbbbb
bbbbbbbbcacaca
bbbbbbbbbbca
bbbcabbbbbbbca
bbbcacacacacacacac
```

(b)

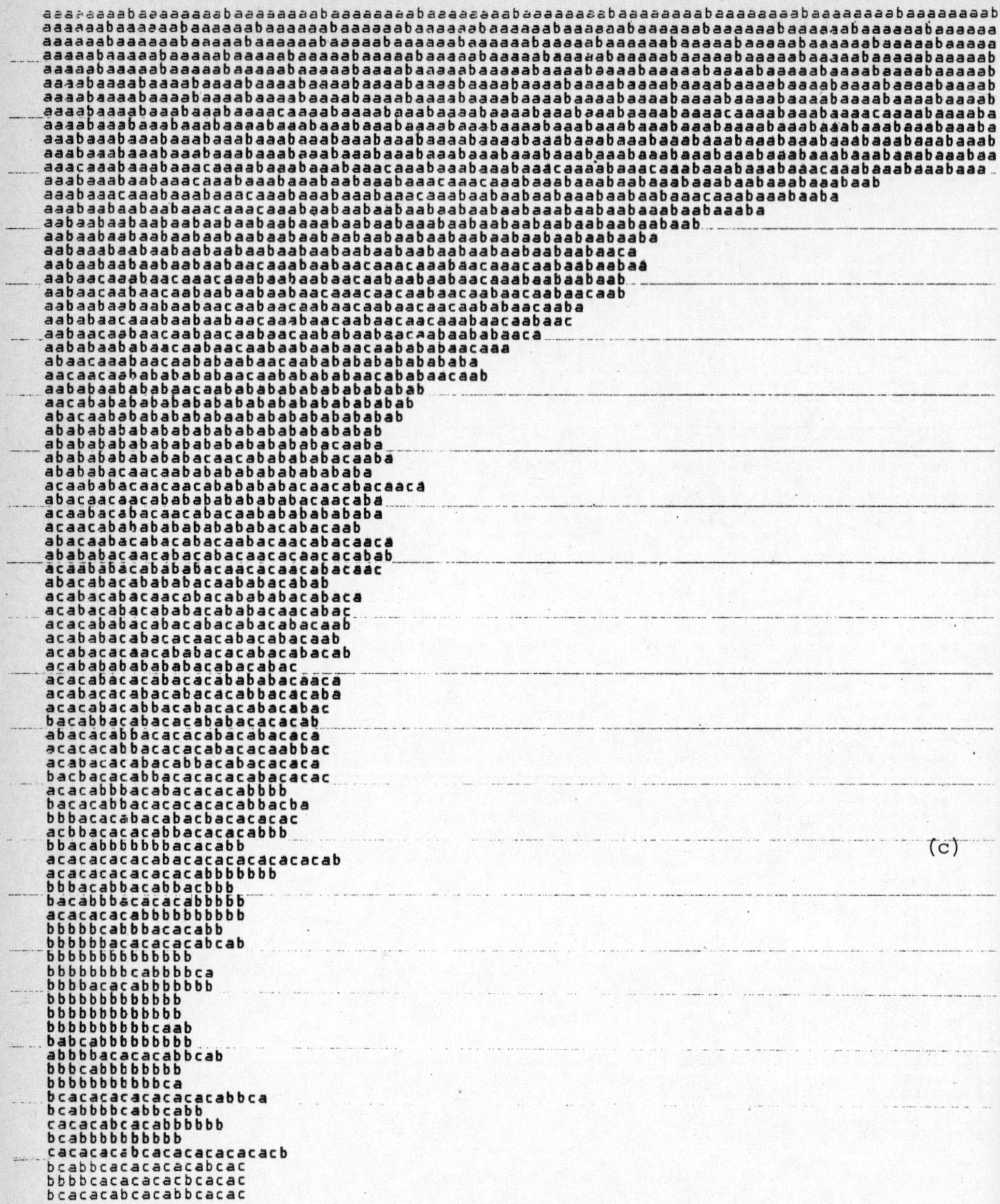

```
aaaaaaaabaaaaaaaabaaaaaaaabaaaaaaaabaaaaaaaabaaaaaaaabaaaaaaaabaaaaaaaabaaaaaaaabaaaaaaaab
aaaaaabaaaaaabaaaaaabaaaaaabaaaaaabaaaaaabaaaaaabaaaaaabaaaaaabaaaaaabaaaaaabaaaaaabaaaaaa
aaaaaabaaaaaabaaaaabaaaaaabaaaaabaaaaaabaaaaaabaaaaabaaaaaabaaaaabaaaaabaaaaaabaaaaabaaaaa
aaaaabaaaaabaaaaabaaaaabaaaaabaaaaabaaaaabaaaaabaaaaabaaaaabaaaaabaaaaabaaaaabaaaaabaaaaab
aaaaabaaaaabaaaaabaaaaabaaaaabaaaabaaaaabaaaaabaaaaabaaaaabaaaabaaaaabaaaabaaaabaaaaab
aaaabaaaabaaaabaaaabaaaabaaaabaaaabaaaabaaaabaaaabaaaabaaaabaaaabaaaabaaaabaaaabaaaabaaaab
aaaabaaaabaaaabaaaabaaaabaaaabaaaabaaaabaaaabaaaabaaaabaaaabaaaabaaaabaaaabaaaabaaaabaaaab
aaaabaaaabaaabaaabaaaacaaaabaaaabaaabaaaabaaaabaaaabaaabaaabaaaabaaaacaaaabaaabaaaacaaaabaaaaba
aaaabaaabaaabaaabaaaabaaabaaabaaabaaaabaaabaaaabaaaabaaabaaabaaabaaabaaaabaaabaaabaaaabaaaba
aaabaaabaaabaaabaaabaaabaaabaaabaaabaaaabaaaabaaabaaabaaabaaabaaabaaabaaabaaabaaabaaabaaab
aaabaaabaaabaaabaaabaaabaaabaaabaaabaaabaaabaaabaaabaaabaaabaaabaaabaaabaaabaaabaaabaaabaa
aaacaaabaaabaaacaaaabaaabaaabaaacaaabaaabaaabaaabaaacaaaabaaacaaaabaaabaaabaaacaaabaaabaaabaaa
aaabaaabaabaaacaaabaaabaaabaabaaabaaacaaacaaabaaabaaabaaabaabaaabaaabaabaaabaaabaab
aaabaaacaaabaaabaaacaaabaaabaaabaaacaaabaabaabaabaaabaabaabaaacaaabaaabaaba
aaabaabaabaabaaacaaacaaabaabaabaabaabaabaabaabaaabaabaaabaabaaaba
aabaabaabaabaabaabaabaabaaabaabaabaaabaabaabaabaabaabaabaab
aabaabaabaabaabaabaabaabaabaabaabaabaabaabaabaabaabaaba
aabaaabaabaabaabaabaabaabaabaabaabaabaabaabaabaabaabaaca
aabaabaabaabaabaacaaabaabaacaaacaaabaacaaacaabaabaabaa
aabaacaaabaacaaacaaabaabaabaacaabaabaabaacaaabaabaabaab
aabaacaabaacaabaabaabaabaacaaacaacaabaacaabaacaabaacaab
aabaabaabaabaabaacaabaacaabaacaabaacaacaababaacaaba
aababaacaaabaabaabaacaaabaacaabaacaacaaabaacaabaac
aabaacaabaacaabaacaabaacaababaabaacaabaababaaca
aababaababaacaabaabaabaacaabababaacaaa
abaacaaabaacaababaabaacaababababababababa
aacaacaababababaacaabababaacababaacaab
aababaabababaacaabababababababababab
aacababababababababababababababab
abacaababababababaabababababababab
ababababababababababababababab
ababababababababababababacaaba
ababababababacaacababababacaaba
abababacaacaababababababababa
acaababacaacaacababababacaacabacaaca
abacaacaacababababababacaacaba
acaabacabacaacabacaabababababa
acaacababababababacabacaab
abacaabacabacabacaabacaacabacaaca
abababacaacabacabacaacacaacacabab
acaababacababababacaacacaacabacaac
abacabacababacaababacabab
acabacabacaacabacababababacabaca
acabacabacabababacababacaacabac
acacababacabacabacabacabacaab
acababacabacacaacabacabacaab
acabacacaacabababacacabacabacab
acababababababacabac
acacabacacabacacabababacaaca
acabacacabacabacacabbacacaba
acacabacabbacabacacabacabac
bacabbacabacacababacacacab
abacacabbacacacabacabacaca
acacacabbacacacabacacaabbac
acabacacabacabbacabacacaca
bacbacacabbacacacacabacacac
acacabbbacabacacacabbbb
bacacabbacacacacacabbacba
bbbacacabacabacbacacacac
acbbacacacabbacacacabbb
bbacabbbbbbbacacabb
acacacacacabacacacacacacacab
acacacacacacacabbbbbbb
bbbacabbacabbacbbb
bacabbbacacacabbbbb
acacacacabbbbbbbbbb
bbbbbcabbbacacabb
bbbbbacacacacabcab
bbbbbbbbbbbbbb
bbbbbbbbcabbbbca
bbbbacacabbbbbbb
bbbbbbbbbbbbb
bbbbbbbbbbbbb
bbbbbbbbbbcaab
babcabbbbbbbbb
abbbbacacacabbcab
bbbcabbbbbbbb
bbbbbbbbbbca
bcacacacacacacacabbca
bcabbbbcabbcabb
cacacabcacabbbbbb
bcabbbbbbbbbb
cacacacabcacacacacacacb
bcabbcacacacacabcac
bbbbcacacacacbcacac
bcacacabcacabbcacac
```

(c)

*** Figure 4.10 - I/O patterns obtained upon submitting the model to stimuli of periods between 5 ms (top) and 430 ms (bottom). (a) σ_{Pb_1}=0 mv. (b) σ_{Pb_1}=1.5 mv. (c) σ_{Pb_1} =3.5 mv.

In Figure 4.11, some of the above I/O patterns are statistically characterized in terms of the partial patterns between discharges that they contain.

Figure 4.12 supposes a greater data reduction, since it represents the average firing period as a function of the stimulation period.

From the above graphics, and other similar ones obtained for several discretization levels and values of σ_{Pb_1}, the following overall trends are deduced:

(a) The ratio between the stimulation period and the firing period becomes more exponential as the dispersion of the random variable Pb_1 increases. This is due to a progressive narrowing of the bands of frequencies that yield entrainment.

(b) The mentioned narrowing is approximately symmetric, despite the tendency of the interspike interval to lengthen with the standard deviation of Pb_1.

(c) The types of partial patterns that appear in the I/O patterns corresponding to the upper frequencies of the entrainemnt bands shift progressively to the left, within the spectrum $\mathbf{a}^n\mathbf{b}$, $\mathbf{a}^n\mathbf{c}$, $\mathbf{a}^{n-1}\mathbf{b}$, $\mathbf{a}^{n-1}\mathbf{c}$, etc., as σ_{Pb_1} increases, the magnitude of the shift being larger for higher frequencies. In the I/O patterns corresponding to the lower frequencies of the entrainment bands, the shift takes place to the right and lengthens as the frequency decreases.

(d) The variance of the histogram of instances of the different partial patterns within an I/O pattern grows with σ_{Pb_1}.

(e) The more **c**'s an I/O pattern contains, the greater distorting effect randomness has on that pattern. Subharmonic entrainments are, therefore, the most stable.

(f) Among the I/O patterns with an equal number of **c**'s, the most sensitive to randomness are those with a higher entrainment ratio. Thus, for example, the lower the subharmonic, the less the observed distortion.

(g) Compensatory effects occur, such as the substitution of a partial pattern $\mathbf{a}^n\mathbf{b}\mathbf{a}^n\mathbf{b}$ for $\mathbf{a}^n\mathbf{c}\mathbf{a}^{n-1}\mathbf{b}$. The analysis of this phenomenon requires a

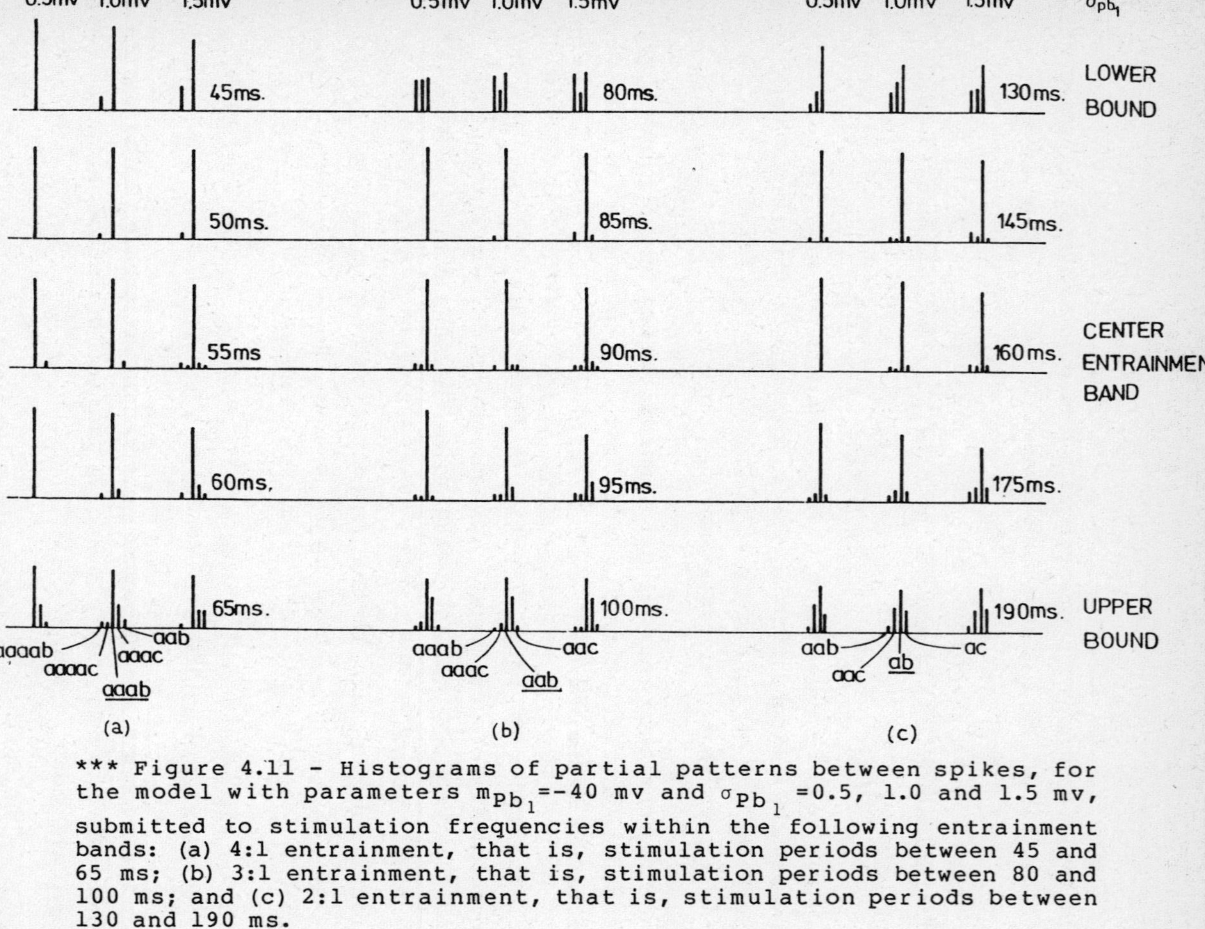

*** Figure 4.11 - Histograms of partial patterns between spikes, for the model with parameters m_{Pb_1}=-40 mv and σ_{Pb_1} =0.5, 1.0 and 1.5 mv, submitted to stimulation frequencies within the following entrainment bands: (a) 4:1 entrainment, that is, stimulation periods between 45 and 65 ms; (b) 3:1 entrainment, that is, stimulation periods between 80 and 100 ms; and (c) 2:1 entrainment, that is, stimulation periods between 130 and 190 ms.

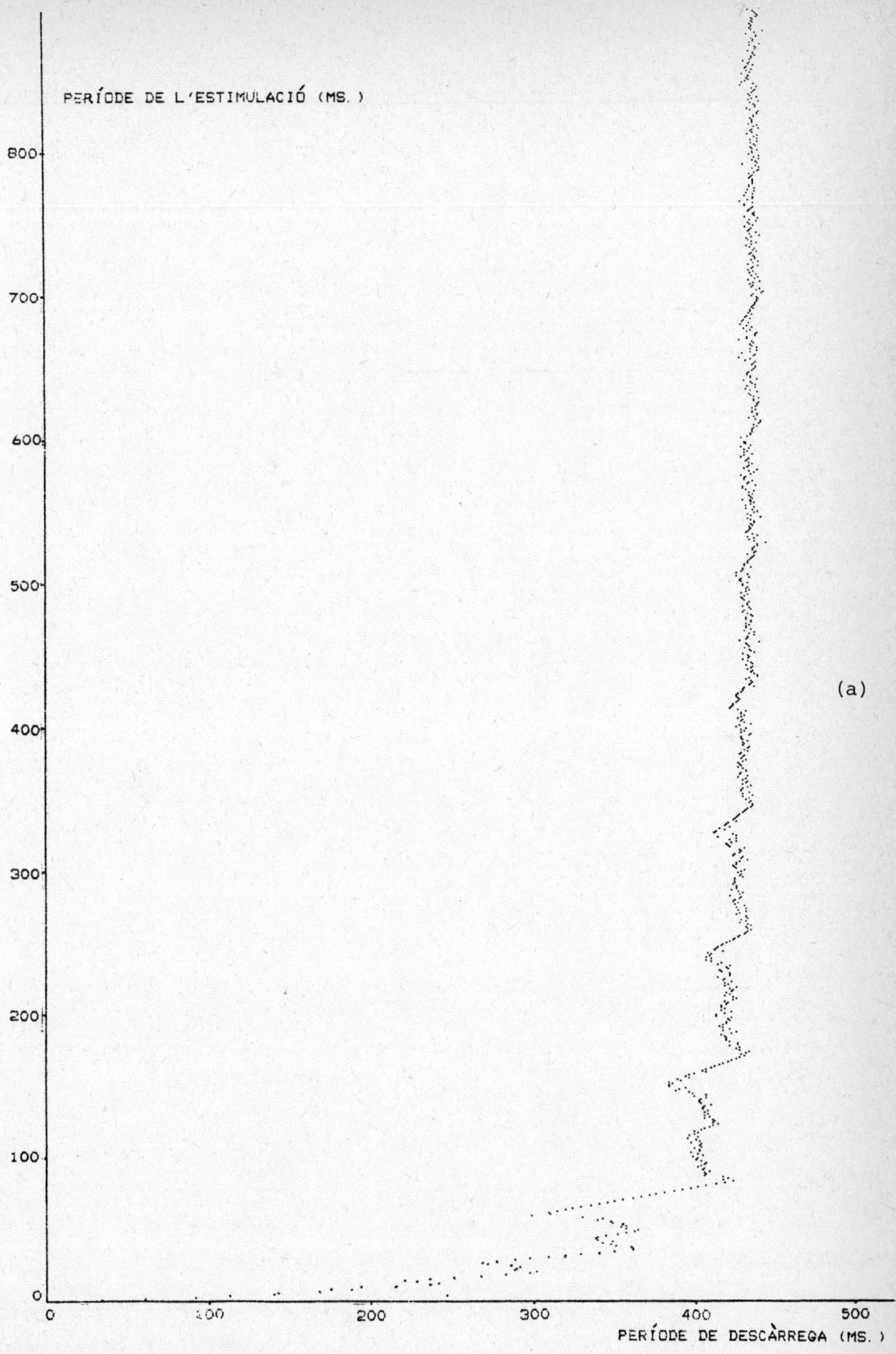
PERÍODE DE L'ESTIMULACIÓ (MS.)
800
700
600
500
400
300
200
100
0
0
100
200
300
400
500
PERÍODE DE DESCÀRREGA (MS.)
(a)

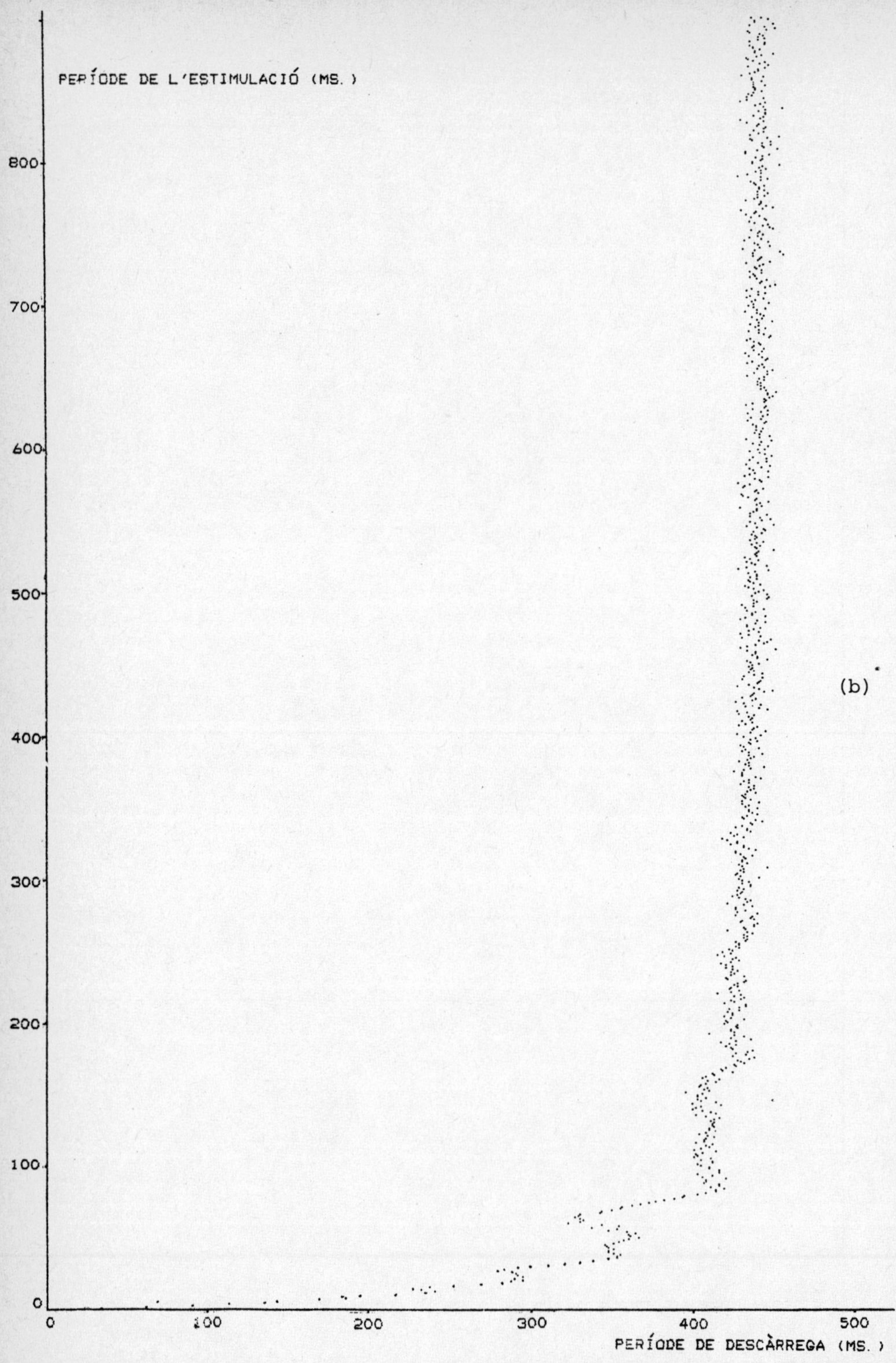

PERÍODE DE L'ESTIMULACIÓ (MS.)
800
700
600
500
400
300
200
100
0
0
100
200
300
400
500
PERÍODE DE DESCÀRREGA (MS.)
(b)

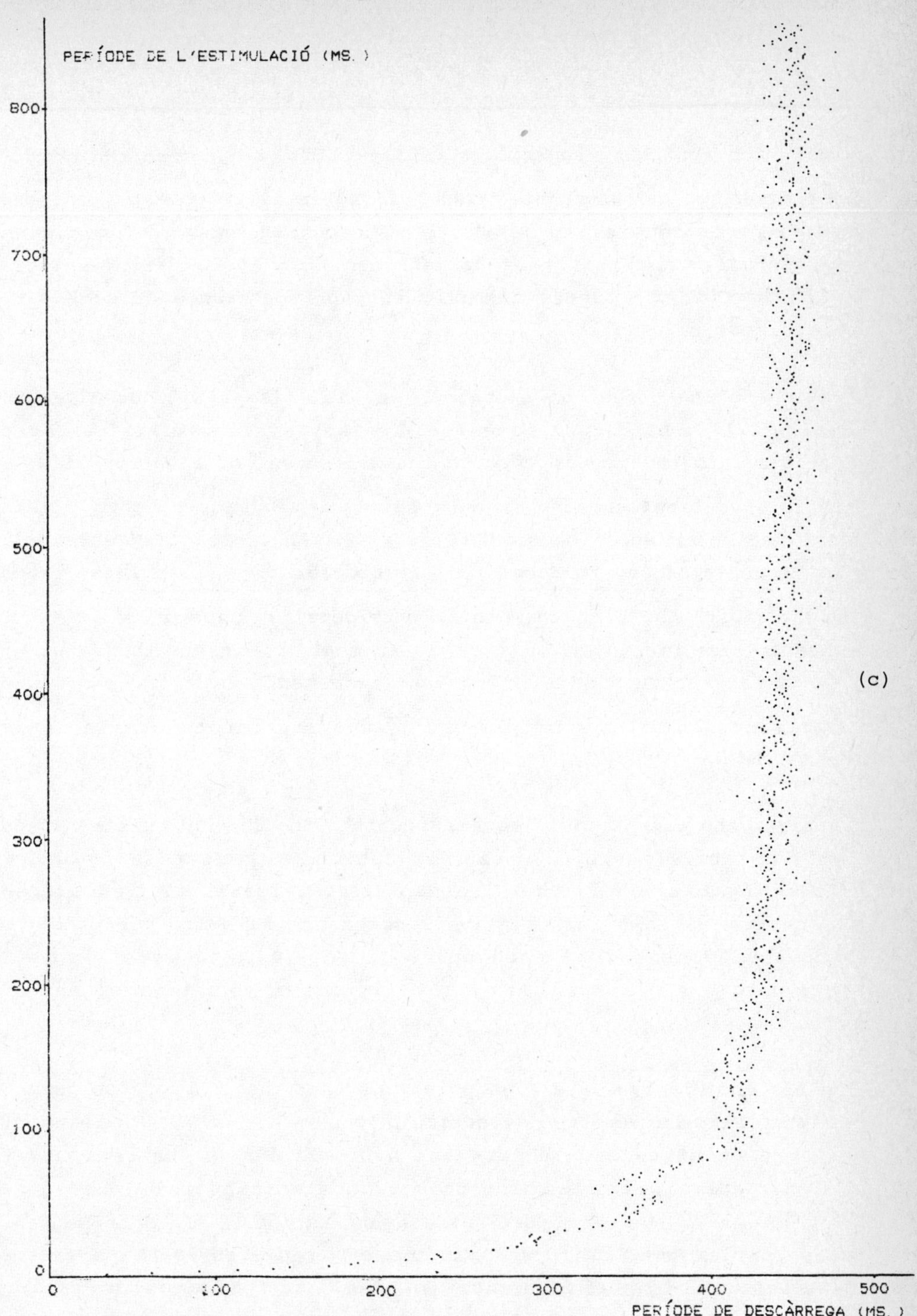

*** Figure 4.12 - Representation of the average inter-spike interval as a function of the period of the stimulus. (a) σ_{Pb_1}=1 mv. (b) σ_{Pb_1}=2 mv. (c) σ_{Pb_1}=3 mv.

statistical study, not only of the partial patterns appearing between consecutive spikes, but also of the partial patterns between every two spikes, three spikes, etc.

We finally mention that, regardless of the value of σ_{Pb_1} and the stimulation frequency, the stochastic model is potentially able to generate all I/O patterns consisting of **a**'s, **b**'s and **c**'s. The gamut of patterns of the stochastic model is thus larger than that of the deterministic model, though for the former the frequency of appearance of some patterns is almost nil.

In relation to the effect of learning, we will also limit ourselves to indicating a line of future research. The way this capacity has been incorporated into the model, through the selective modification of m_{Pb_1}, permits interpreting learning as a directional tendency to move within the (N,λ)-parameter space shown in Figure 4.7. An increment (decrement) in m_{Pb_1} --corresponding to accelerative (decelerative) learning-- leads to a right (left) shift in the probability density function of Pb_1 and, consequently (see Figure 4.9), tends to augment (diminish) the value of λ. Simultaneously, that same increment (decrement) in m_{Pb_1} determines also a diminution (augmentation) in T_0 and, therefore, an increase (decrease) in N.

Summarizing, the direction of movement within the (N,λ)-parameter space is always positively-sloped, its orientation being increasing or decreasing, depending on whether learning is decelerative or accelerative. Figure 4.13 shows the relative lengths of the vectors expressing the tendency to move from each point of the (N,λ)-parameter space, together with the demarcation of the regions of entrainment and of learning.

At a purely speculative level, we point out that, because of the geometric relationship between the direction of movement and the entrainment regions, learning tends to carry the neuron towards the regions of subharmonic entrainment --which thus become attractors-- therefore favoring the emergence of simple entrainment ratios to the detriment of the more complex ones. The predominance of simple entrainment ratios has been reported by numerous experimenters, as indicated in Section 2.1.1.3. A similar reasoning points out the tendency of randomness to make the patterns vary within only one entrainment region, rather than make them jump from one region to another, as would happen in the case that the variation of Pb_1 took place in the direction perpendicular to the current one.

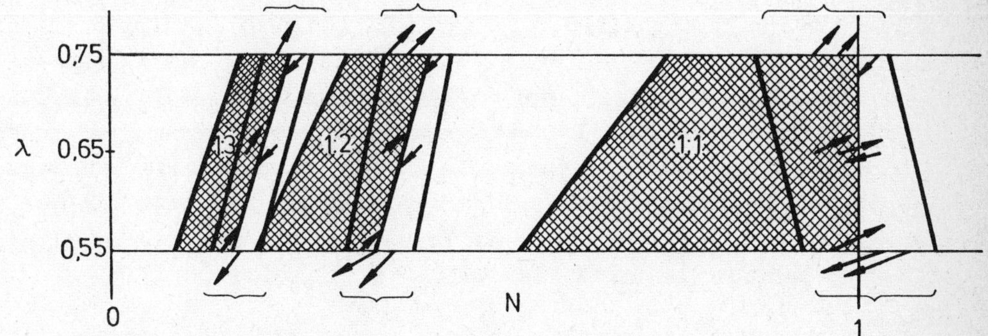

*** Figure 4.13 - Vectors that characterize learning in the different regions of the (N,λ)-parameter space where it takes place (between brackets), in relation to the entrainment areas (shaded).

4.5 - CONCLUSIONS

"Good problems and mushrooms of certain kinds have something in common: they grow in clusters. Having found one, you should look around: there is a good chance that there are some more quite near".

H. Polya, 1945

If the conditions stated in Section 2.1.1.2 hold, the behavior of an ideal oscillator submitted to a periodic input is uniquely determined by its response to occasional perturbations. Thus, to study the regions and patterns of entrainment, a detailed mechanistic model is not required, but a phenomenological model, such as the PRC, suffices.

In the present chapter we have analyzed the entrainment consequences of having a particular piecewise linear PRC (Figure 4.1), which approximates the one obtained in the preceding chapter for the model there proposed and even presents the more general interest of having appeared --as a result of carrying out a linear regression on the data-- in a great number of experimental preparations, consisting of a pacemaker neuron with an excitatory synapse, presumably obeying different underlying mechanisms.

The regions in the space of initial phases and relative stimulation frequencies where the different transitions take place and the infinite possible entrainment ratios are attained have been determined, the conclusion being that the I/O pattern is periodic except for a set of input frequencies of measure zero. Thus, if this deterministic phenomenological model (the PRC) appropriately described the biological reality, aperiodic patterns would never appear and it would only be necessary to wait long enough for the cycle to be closed. However, because of intrinsic randomness in neural pacemakers, the situation is quite the opposite, it being only necessary to wait long enough for a pattern that seemed periodic to break its cycle.

The complete characterization of the I/O patterns corresponding to entrainment attained constitutes a basic framework to test the adequacy of the different hypotheses introduced with respect to a specific biological preparation.

In particular, if there is available an experimental PRC of the type studied, as well as the I/O patterns corresponding to several stimulation frequencies and a dispersion measurement of the histogram of spontaneous interspike intervals, then some statistical tests based on the elaboration of the data sketched in Section 4.4 can be applied to elucidate if the mentioned dispersion justifies the observed divergences from the deterministic patterns.

Two reasons can account for a failing in the above tests: an erroneous incorporation of the source of randomness into the model (the effect of randomness can differ, for example, in the presence or absence of stimulation) or the inadequacy of the hypotheses that permit deriving the phase transition equation (the conditions described in Section 2.1.1.2 may not hold or too much information may be lost in the piecewise linear approximation of the PRC).

For comparative purposes, and taking into account its being obtained experimentally through regression, it could be of interest to analyze the implications of introducing slight modifications in the piecewise linear PRC studied, such as for example admitting that $\delta(0)$ and $\delta(1)$

could adopt values different from zero or that the slope of the segment corresponding to the upper phases could differ from 1. The modified PRCs would give place to phase transition curves on the circle discontinuous at the origin, of rotation number greater than 1 or decreasing in certain segments.

Curves with such characteristics are the subject of intense research (Li and Yorke, 1975; May, 1976; Guckenheimer, 1977; Štefan, 1977; Keener, 1980), it having been proven that some among them generate, in a deterministic way, aperiodic patterns for ranges of stimulation frequencies of positive measure.

The fact that this result holds both for PRCs with a discontinuity (Keener, 1980) and for PRCs continuous at least up to the second derivative (Herman, 1977), but not for the type of PRC we have considered --which, being continuous and having a discontinuity in its first derivative, falls somewhere in between-- raises the theoretical question of what the key aspect is that makes PRCs such as ours so special.

Despite the undeniable incidence of randomness in the I/O patterns of pacemaker neurons, it would be interesting to elucidate if other causes, such as the ones mentioned above involving slightly modified PRCs, could also account for the variations observed in the mentioned patterns and, if this is the case, in what proportion they depend on one or the other cause.

In regard to the type of learning approached, we have already indicated that, for the model proposed in the preceding chapter, it implies a consistent modification of the PRC, which permits taking advantage of the complete division of the (N,λ)-parameter space into entrainment regions, to study the regions that act as attractors.

In sum, the results obtained in the present chapter answer a series of questions previously raised by Segundo and Kohn (1981), relative to the patterns of entrainment between a pacemaker neuron, characterized by a piecewise linear PRC, and a periodic input; while at the same time suggesting new questions, especially in relation to the effect of randomness and learning upon the mentioned patterns.

CHAPTER 5

MODELLING AND SIMULATION OF A NETWORK OF PLASTIC PACEMAKER NEURONS

"What is simple is not true, what is complex is useless".

P. Valéry

In Section 2.1.2 evidence was presented that certain nervous structures of vertebrate animals can assimilate and later autonomously reproduce rhythms previously received through a sensory modality. For a neuronal network to incorporate this capacity, it must be provided with at least two mechanisms: one of endogenous generation of rhythms and another of modification of its firing pattern. Both of these mechanisms can be in the individual neurons or be a property emerging from the connectivity of the network. The number of possibilities increases when realising that, between the individual and the collective extremes, there is a wide spectrum of hybrid systems in which the mentioned mechanisms act jointly at both levels.

In the present work, we have chosen to explore the implications of incorporating autorhythmicity and plasticity at the level of the individual neuron. The experimental plausibility of this choice in invertebrates, with regard to plasticity, has been discussed in Section 3.2.2. Complementary support for the existence of pacemaker neurons in vertebrates, and for the similarity of their functioning with those of invertebrates, will be furnished in Section 5.1.2. The evidence given supports the interest of the chosen option, but does not disqualify other options. The experimental data in this field are varied enough to justify the exploration of different possibilities. Therefore, more than just for the data, the choice of option is motivated by the theoretical interest of establishing different alternatives, in the sense expressed by von Baumgarten (1970) in relation to plasticity[£] and by

----

(£): "No major breakthrough in the synaptic connectivity hypothesis of memory has been achieved, despite extensive work during the past 10 years, which justifies our focussing attention on sites and possibilities other than the synapses in which information may be stored" (von Baumgarten, 1970).

Hartline (1979), in the domain of autorhythmicity[(£)].

The objective that we intend to achieve with the study of the proposed network is to determine under which conditions it can serve as a substratum to the phenomena of assimilation and reproduction of rhythms previously described and to analyze the requirements that these behaviors impose upon the parameters that define the connectivity, the functioning of each neuron and the initial conditions. Actually, the goal is to obtain better knowledge of the relevant structural and functional factors at the microscopic level for the appearance of the mentioned macroscopic phenomena. It is important to make clear that we will only touch on the aspects of assimilation and reproduction of rhythms, without entering into the question of their behavioral significance; in other words, into the mechanisms that determine their association with other stimuli or specific behaviors.

The present chapter is structured as follows: The first section describes the proposed network model and provides the evidence to which we referred above. The second section specifies certain details of implementation, as well as the type of measures and graphics used to characterize the behavior of the network. The third section reports and comments on the results of simulation, establishing in the first place a reference experiment, from which the effects that variation of different factors has upon the behavior of the network is analyzed. Finally, the conclusions that follow from the results obtained are outlined and possible lines of future research are indicated.

5.1 - THE NEURAL NETWORK MODEL PROPOSED

The network model we propose does not attempt to be a faithful replica of any specific nervous structure, but does, however, adhere to certain structural and qualitative principles attributed to the thalamus and to the sensory areas of the cortex.

We will first describe the internal connectivity and the inputs of the network, then give experimental support for the chosen options and, finally, provide a formal specification of the network model proposed.

(£): "...there has been a persistent tendency of network physiologists to look primarily to the connectivity pattern among neurons as an explanation for network behavior. In fact, equal attention should be paid to functional dynamics of individual cells and synapses" (Hartline, 1979).

5.1.1 - Description of the connectivity

We will consider networks of 10 x 10 pacemaker neurons, situated on the nodes of a toroidal network with square lattice, where the connectivity grossly obeys the architectonic principle of feedback lateral inhibition, being random at the microlevel. Thus, each neuron establishes excitatory connections with its closest neighbors and inhibitory connections --probably mediated by interneurons, which we do not explicitely model-- with others further away, according to a specific probability distribution. All connections have unit weight, but input amplitudes vary due to changes in the output amplitude of presynaptic neurons.

Figure 5.1 exemplifies the principle of feedback lateral inhibition, contrasting it with that of feedforward lateral inhibition.

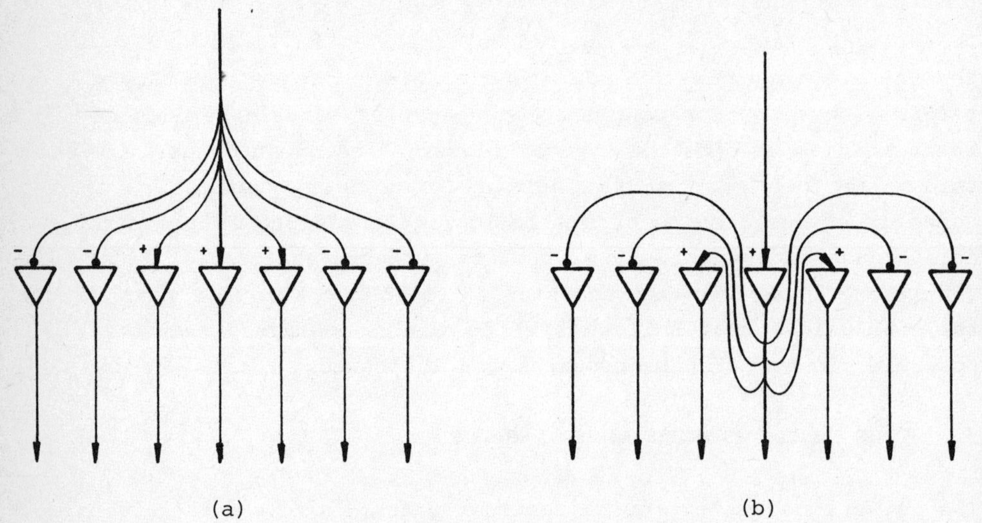

*** Figure 5.1 - Exemplification of the lateral inhibition principle, in its two versions: (a) feedforward; (b) feedback. + = excitatory synapse, - = inhibitory synapse.

The probability that two neurons are connected excitatorily or inhibitorily, depends on the distance (d) between them and not on their location or their relative orientation in the network. As usual in the specification of connectivities of this kind --called "homogeneous"-- we will start by defining the one-dimensional probability density function that characterizes the establishment of connections at different distances, and then, taking into account the metric of the network, will derive the two-dimensional probability density function of a neuron establishing connection with each other neuron of the network(£). We will divide each of the mentioned probability density functions into two: one for the excitatory connections and another for the inhibitory ones.

Figure 5.2 shows the two one-dimensional probability density functions that characterize the establishment of excitatory and inhibitory connections of different lengths starting from a neuron. The function corresponding to excitatory connections $p_e(d)$ is represented in the positive direction of the axis of ordinates and the one corresponding to the inhibitory ones $p_i(d)$, in the negative direction of the same axis. The two parameters that characterize them are: the range of excitation (a_e) and the range of inhibition (a_i), both measured taking as a unit the separation between consecutive neurons in the directions of the coordinate axes.

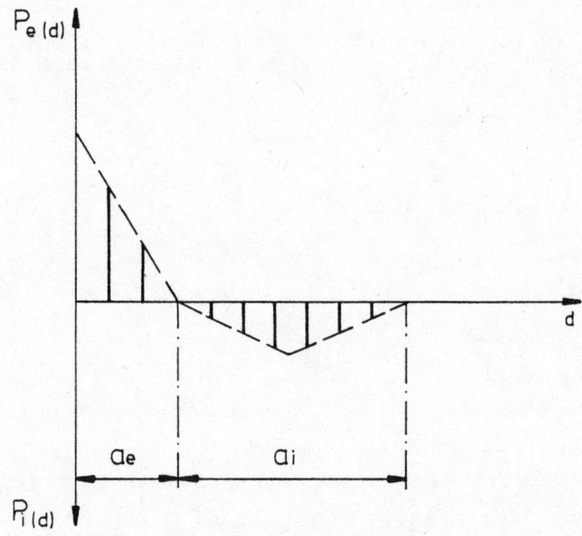

*** Figure 5.2 - Probability density functions that characterize the establishment of excitatory connections --$p_e(d)$-- and inhibitory connections --$p_i(d)$-- at various distances (d).

(£): In the case of a continuous space with Euclidean metric, the second density function is obtained by rotating the first one around the axis of ordinates, discounting a factor of normalization.

The equations that define the two mentioned one-dimensional functions are:

$$p_e(d) = \begin{cases} \dfrac{a_e - d}{\sum_{i=1}^{a_e - 1} i}, & \text{if} \quad 0 < d < a_e \\ 0, & \text{otherwise} \end{cases} \tag{5.1}$$

$$p_i(d) = \begin{cases} \dfrac{\min(d-a_e,\ a_e+a_i-d)}{\sum_{i=1}^{[a_i/2]} i + \sum_{i=[ai/2]+1}^{a_i} (a_i - i)}, & \text{if} \quad a_e < d < a_e + a_i \\ 0, & \text{otherwise} \end{cases} \tag{5.2}$$

where d is measured in the same units of separation between neurons as a_e and a_i.

It is clear that, defined this way, $p_e(d)$ and $p_i(d)$ are actually probability density functions, since

$$\sum_{d=-\infty}^{\infty} p_e(d) = \sum_{d=-\infty}^{\infty} p_i(d) = 1 \tag{5.3}$$

To derive the two-dimensional probability density functions that govern the establishment of connections between one neuron and each other neuron of the network, we will use the metric induced by the block-distance, which between any point (x,y) of the plane and the origin is given by:

$$d(x,y) = |x| + |y|$$

According to this metric, the number of neurons at a distance d from a fixed neuron is 4d. Consequently, we deduce from equations 5.1 and 5.2 the following expressions for the two-dimensional functions:

$$p_{2e}(x,y) = \frac{p_e(|x|+|y|)}{4(|x|+|y|)} = \begin{cases} \dfrac{a_e - |x| - |y|}{4(|x|+|y|)\sum_{i=1}^{a_e-1} i}, & \text{if } 0 < |x|+|y| < a_e \\ 0, & \text{otherwise} \end{cases} \qquad (5.4)$$

$$p_{2i}(x,y) = \frac{p_i(|x|+|y|)}{4(|x|+|y|)} = \begin{cases} \dfrac{\min(|x|+|y| - a_e,\ a_e+a_i-|x|-|y|)}{4(|x|+|y|)\left[\sum_{i=1}^{[a_i/2]} i + \sum_{i=[a_i/2]+1}^{a_i} (a_i-i)\right]}, & \text{if } a_e < |x|+|y| < a_e + a_i \\ 0, & \text{otherwise} \end{cases} \qquad (5.5)$$

It is easy to demonstrate that these functions fulfill the two-dimensional version of the property expressed in equation 5.3.

In Figure 5.3 we find the above two-dimensional functions graphically represented. Considering that they only depend on one variable (d), the equiprobability lines coincide with the lines equidistant from the origin.

Together with the distributions $p_{2e}(x,y)$ and $p_{2i}(x,y)$, two other factors characterize the connectivity of the network: the number of connections per neuron (n) and the proportion of excitatory connections over the total (p).

The effects that variation of these parameters has upon the behavior of the network, as well as those of the introduction of slight modifications in the connectivity pattern described, will be dealt with in Section 5.3.2.1.

Finally, in relation to the introduction of stimulation, all the neurons of the network incorporate an excitatory external input; in the present work only situations in which the same rhythmic stimulus is applied simultaneously to all the neurons of the network will be explored.

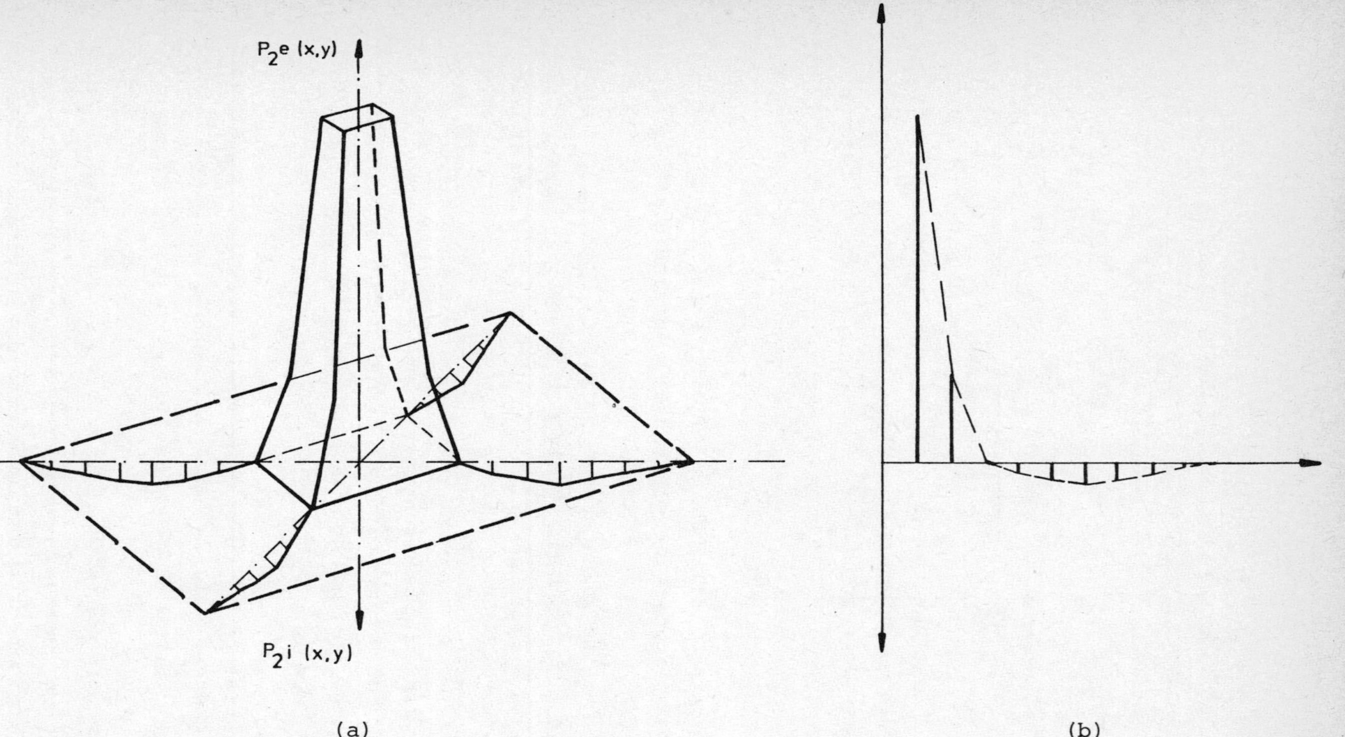

*** Figure 5.3 - (a) Shape of the two-dimensional probability density functions that govern the establishment of excitatory connections ($p_{2e}(x,y)$) and inhibitory connections ($p_{2i}(x,y)$), between neurons of the network. (b) A slice parallel to the probability axis of the mentioned functions, showing the discretization disregarded in the other graph, as well as the relative probability in each of the equiprobability lines.

5.1.2 - Experimental support

As said in the introduction, the evidence that will be given in this section supports the interest of exploring the proposed network model, but in no way its exclusivity, since it does not invalidate alternative equally plausible options.

The aspects that we will be concerned with are basically of two kinds: those relative to the existence of pacemaker neurons in vertebrates and those characterizing the connectivity in sensory nervous structures.

The final demonstration that the rhythmic firing pattern of a neuron is originated by intrinsic processes and does not derive from external stimulation requires recording the electrical activity of the mentioned neuron when all its synaptic connections have been removed. This method, whose application to nervous systems of invertebrates does not entail serious difficulties, is, on the contrary, practically unworkable in the case of vertebrates. Still, experimental results have been obtained which would be difficult to explain without admitting the existence of pacemaker neurons in certain nervous structures of vertebrates. We mention, for example, the extremely regular interspike interval shown by some neurons of the anterior brain of the cat (Smith and Smith, 1964, 1965) and the hypothalamus of the rabbit (Vinogradova, 1970), when submitted to a variety of stimulation conditions; the permanent phase-shift in the oscillatory activity of fusimotor neurons, provoked by an inhibitory impulse (Fidone and Preston, 1971); the systematic variation of the firing frequency of certain abdominal motor neurons again of the cat, when injected with constant depolarization currents (Schwindt and Calvin, 1973; Calvin, 1974); and other similar data obtained from neurons that generate the breathing rhythm in different mammals (Wyman, 1977; Feldman and Cleland, 1982) and from the neurons of the supraoptic area of the hypothalamus of the rat (Gahwiler and Dreifuss, 1979).

Several researchers, on the basis of very diverse experimental results, have pointed out the interest of the neurobiological studies of invertebrates for discovering the principles of neuronal functioning in vertebrates (Strumwasser, 1967; Kandel and Spencer, 1968; Bullock, 1976; Selverston, 1976; Barker and Smith, 1978; Ayers and Selverston, 1979; Kohn et al, 1981). Comparative studies of neurons belonging to vertebrates and invertebrates have been carried out by Cohen (1970) and

Bullock (1974). In the specific field of plasticity, Kandel (1976) notes that the analyses at the cellular level of habituation in the nervous systems of vertebrates are consistent with the picture that had previously emerged from the studies on Aplysia and crayfish; and that, specifically, of the nine aspects that characterize short-term habituation in vertebrates, eight had already been observed in Aplysia. The compilation of articles about neural learning edited by Teyler (1978) offers a comparative view of the mechanisms used at different levels by vertebrates and invertebrates. Summarizing, it seems that there is consensus that many strategies of modification at the unineuronal level may coincide in both, though major differences may occur at the level where these modifications are integrated to yield a change in the behavior of the network, because of its dependence on the structure of the network and on the strategies used by the organism (Arbib et al., 1976).

With regard to the connectivity of the network, lateral inhibition, in its two modalities --feedforward and feedback-- seems to be involved in the dynamics of almost all brain structures. The first modality predominates in the most peripheral parts of the sensory systems and causes an increase of the contrast, since it concentrates the excitation within very limited areas; its most characteristic example is the retina (Hubel and Wiesel, 1959, 1963; Lettvin et al., 1960). Feedback inhibition has basically three effects: the synchronization of a group of neurons, the segregation in time of certain events and the bounding of excitation; structures that incorporate this modality are the spinal motoneuronal loop that contains the Renshaw cells (Eccles et al., 1954; Brooks and Wilson, 1959; Szentágothai and Arbib, 1974), the thalamic regions through which the sensory projections go toward the cortex (Andersen and Eccles, 1962; Andersen et al., 1964; Thatcher and Purpura, 1972, 1973; Singer, 1977) and also, presumably, the cortical areas underlying the generation of the alpha rhythm and the EEG (Creutzfeldt et al., 1966; Andersen and Andersson, 1968; Thatcher and John, 1977). In the experiments done with the tracer technique, the microrecordings obtained from the sensory areas of the cortex often show also reciprocally inhibitory connections between two neurons (Ramos et al., 1976a).

We point out that the majority of models of structures involved in the processing of sensory information incorporate feedback lateral inhibition in the design of their connectivity (Hartline et al., 1961; Didday, 1970, 1976; Spinelli, 1970; von der Malsburg, 1973; Arbib et al., 1974; Dev, 1975; Amari, 1977c).

Concerning the stimulation conditions, it is important to highlight that the spatial distribution of information is preserved by the projections from the different sensors to their corresponding areas in the sensory cortex, which is why they are called retinotopic, tonotopic, or somatotopic (Szentágothai and Arbib, 1974; Singer, 1977). This fact determines that neighboring neurons receive similar stimulation.

5.1.3 - **Formal specification**

The formalization of the network model will follow the guidelines marked by Zeigler (1976), in the same line in which the neuron model has been specified in Section 3.2.3. Thus, the discrete-time neural network will be represented by the quadruple (I, N, S, τ), where:

$I = \{0, 1, \ldots 9\} \times \{0, 1, \ldots 9\}$ is the set of indexes,

$N = \{(X, E_{\bar{i}}, O_{\bar{i}}, \delta_{\bar{i}}, \lambda_{\bar{i}}) : \bar{i} \in I\}$ is the set of neurons, characterized as sequential machines according to expressions 3.7,

$S = \{(\bar{s}_{\bar{i},1}, \bar{s}_{\bar{i},2}, \ldots \bar{s}_{\bar{i},n}) \in I^n : \bar{i} \in I\}$ is the set of synapses, and

$\tau = \{\tau_{\bar{i}} : \bar{i} \in I\}$ is a family of mappings that define the transmission of impulses between neurons, in the following way:

$$\tau_{\bar{i}} : O_{\bar{s}_{\bar{i},1}} \times O_{\bar{s}_{\bar{i},2}} \times \ldots O_{\bar{s}_{\bar{i},n}} \times E_{\bar{i}} \longrightarrow E_{\bar{i}} \tag{5.6}$$

$$(o_1, o_2, \ldots, o_n, T, Ps, Pb_1, m_{Pb_1}) \longmapsto (T, Ps + \sum_{j=1}^{n} \mu_j o_j, Pb_1, m_{Pb_1})$$

with:

$$\mu_j = \begin{cases} 1, & \text{if } d(\bar{i}, \bar{s}_{\bar{i},j}) < a_e \\ -1, & \text{if } a_e \leqslant d(\bar{i}, \bar{s}_{\bar{i},j}) < a_e + a_i \end{cases}$$

and being d the block-distance:

$$d((i_1, i_2), (s_1\ s_2)) = |i_1 - s_1| + |i_2 - s_2|$$

In the above specification, synapses S are generated according to the probability density functions $p_{2e}(x,y)$ and $p_{2i}(x,y)$ (equations 5.4 and 5.5) and to the parameter p, described in Section 5.1.1; in other words, if $\bar{I} = (i_1, i_2)$,

$$\text{Prob } (\bar{s}_{\bar{I},j} = (s_1,s_2)) = \begin{cases} p \cdot p_{2e}(|i_1-s_1| \, , \, |i_2-s_2|), & \text{if } |i_1-s_1| + |i_2-s_2| < a_e \\ (1-p)\, p_{2i}(|i_1-s_1| \, , \, |i_2-s_2|), & \text{if } a_e \leq |i_1-s_1| + |i_2-s_2| < a_i \end{cases}$$

The network described can also be represented as a unique sequential machine (X, E, O, δ, λ), where

$X = \mathbb{R}$ is the set of inputs,

$E = \prod_{\bar{I} \in I} E_{\bar{I}}$ is the set of states,

$O = \prod_{\bar{I} \in I} O_{\bar{I}}$ is the set of outputs,

$\delta : E \times X \dashrightarrow E$ is the transition function, indirectly defined:

$$\delta(e_{00}, e_{01}, \dots e_{99}, x) = (e'_{00}, e'_{01}, \dots e'_{99}) \qquad (5.7)$$

so that:

$$e'_{i_1 i_2} = \delta_{\bar{I}} \, (\tau_{\bar{I}}(\lambda_{\bar{s}_{\bar{I},1}} (e_{\bar{s}_{\bar{I},1}}), \dots \lambda_{\bar{s}_{\bar{I},n}} (e_{\bar{s}_{\bar{I},n}}), e_{\bar{I}}))$$

for $\bar{I}=(i_1,i_2)$

$\lambda : E \dashrightarrow O$ is the output function, also defined indirectly:

$$\lambda(e_{00}, e_{01}, \dots e_{99}) = (o_{00}, o_{01}, \dots o_{99}) \qquad (5.8)$$

so that:

$$o_{i_1 i_2} = \lambda_{\bar{I}}(e_{i_1 i_2})$$

for $\bar{I}=(i_1,i_2)$

Table 5.1 provides a listing of the variables and constants that, together with those enumerated in Table 3.1, characterize the network model, indicating at the same time their neurophysiological meaning.

VARIABLES

$\bar{I}$: Index reflecting the position of each neuron in the network.

$\bar{s}_{\bar{I},j}$: Index corresponding to the jth neuron that has established synaptic contact with the neuron of index $\bar{I}$.

CONSTANTS

n : Number of connections per neuron.

p : Proportion of excitatory connections over the total.

a_e : Range of the excitatory connectivity coming out from a neuron.

a_i : Range of the inhibitory connectivity coming out from a neuron.

*** Table 5.1 - List of the variables and constants that characterize the neural network model.

5.2 - IMPLEMENTATION AND POST-PROCESSING TECHNIQUES

The network model was first implemented in a VAX 11/780 at the University of Massachusetts, Amherst, and afterwards extended to the point of configuring a menu-driven system, that includes a set of programs for selecting the experiment to be performed and for processing and graphically representing the results, in the SEL 32/77 of the Institut de Cibernètica at Barcelona.

The basic criterion that has guided the implementation of the model has been to maximally facilitate the carrying out of simulations with different connectivities and values of the parameters that determine the inter- and intra-neuronal functioning. The processes of obtaining and elaborating the results have been separated to favor their optimum exploitation. Figure 5.4 shows schematically the interconnection between the different software modules developed, as well as the data files they require and generate. The three parts that constitute the system: initialization, processing (dynamic evolution) and post-processing of the results, will be briefly described below.

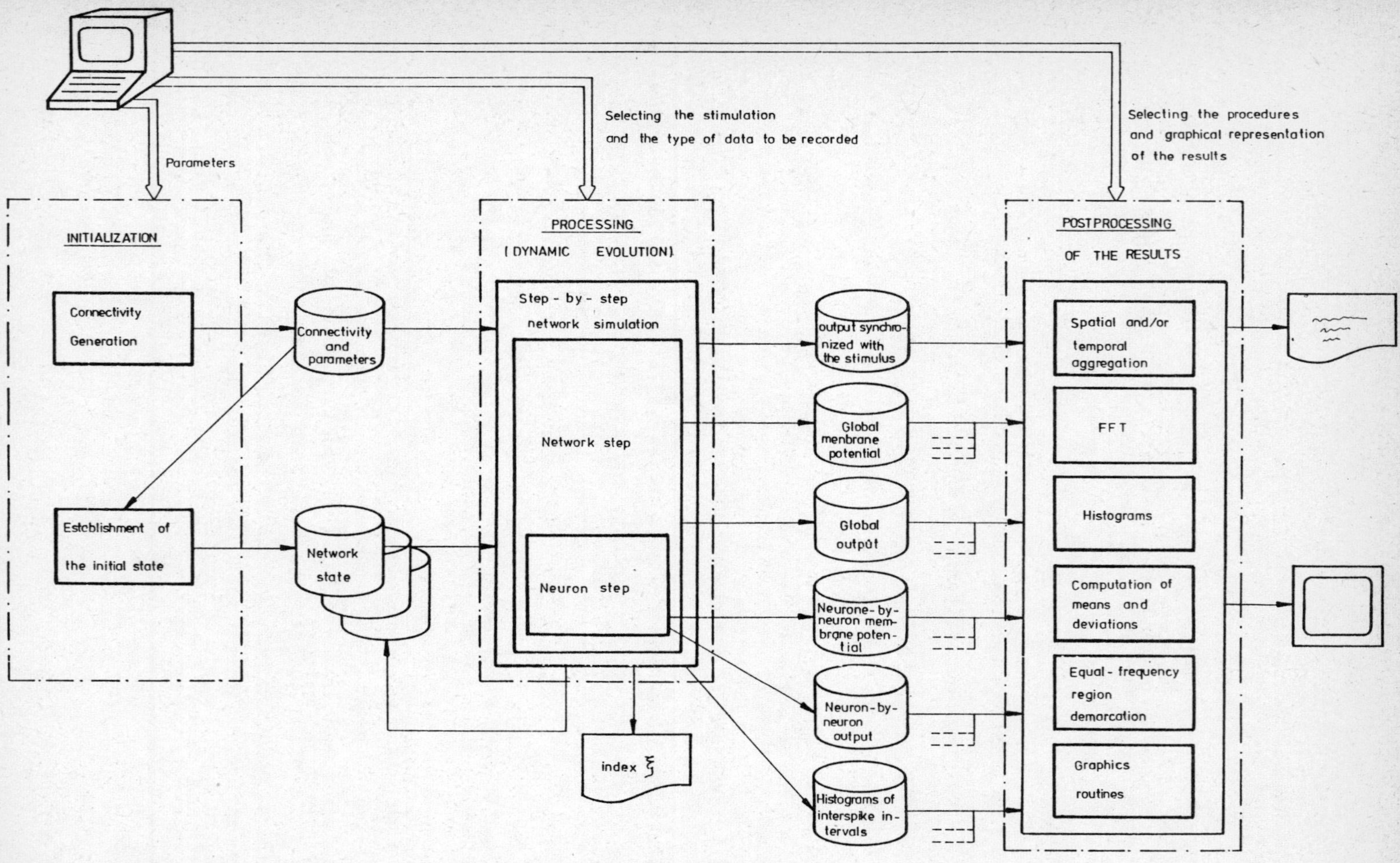

*** Figure 5.4 - Software system developed.

5.2.1 - Initialization

This part contains the modules for generating the connectivity and for establishing the initial state of the neurons in the network. In relation to the first module, once values have been given to the parameters n, p, a_e and a_i, each of the n synapses of each neuron is determined through a double step. First, the distance d to which the presynaptic neuron has to be placed with respect to the postsynaptic one is set, according to the probability distribution function:

$$\text{Prob}(d \leqslant x) = p.\ p_e(x) + (1-p)\ p_i(x) \tag{5.9}$$

where $p_e(x)$ and $p_i(x)$ are given by equations 5.1 and 5.2, respectively. Second, the coordinates of the presynaptic neuron relative to the postsynaptic one, are fixed among the 4d possible pairs, through a uniform probability distribution defined on the interval $[0,4d]$.

Regarding the initial state, characterized by the values of the variables T, Ps, Pb_1 and m_{Pb_1} of each neuron, a two-stage procedure has been used to establish it. The first stage consists in assigning random values according to uniform probability distributions defined in their respective ranges to T, Ps and m_{Pb_1}, and according to a Gaussian distribution with mean m_{Pb_1}, to Pb_1. The second stage has as its objective to generate an initial state compatible with the connectivity of the network. This compatibility has to be understood at two levels: in the short term, the values of T and Ps, for each neuron, must reflect the stimulation coming from other neurons received in the period immediately before. In the longer term, taking into account the neuron's capacity to adapt its own firing rhythm to temporal patterns of activity imposed upon it, the value of m_{Pb_1} has to be consistent with its history of activity during the developmental period in which, without external inputs, it was only excited or inhibited by other neurons of the network.

To attain the aforementioned compatible initial state, the network with a random initial state is submitted to a "running in" process, in which it evolves deprived of external stimulation, the only inputs being the synaptic influences coming from other neurons in the network. Given the wide variations that the intrinsic frequencies of neurons should be allowed to undergo, we have opted for using the second learning rule described in Section 3.2.1.

The implementation of the second stage for the establishment of the initial state, which we have just described, requires to make use of the simulator. This fact, however, has not been reflected in the diagram of Figure 5.4 in order not to obscure the conceptual delimitation of the three parts of the system. The repercussion of the running in process upon the subsequent functioning of the network will be analyzed in Section 5.3.2.3.

5.2.2 - Processing

After the operator has selected the stimulation and the kinds of data to be recorded, the simulator conducts the network evolution during the specified time period. A subroutine, which implements the transition and output functions expressed in equations 5.7 and 5.8, determines the evolution of the network from instant t to instant t+h. The neurons' outputs are not updated until their respective states have been computed. Since the transition and output functions of all neurons (equation 3.7) are identical, one subroutine takes care of the evolution of all.

Each type of result --membrane potential or discharge, individual or global, punctuate or aggregated in time-- is recorded at the level where it is produced. Both the global measurements and those aggregated in time are always obtained through addition. During each period of network evolution specified, the operator has access to an **index ξ** of the evolution of the learning parameters m_{Pb_1}, consisting in the norm of the difference vector of those parameters values for each neuron, in successive uniformly-spaced time instants. Depending on the values taken by this index ξ, the operator can cancel or extend the learning process.

5.2.3 - Post-processing of the results

A set of techniques of discrete-signal processing and graphical representation has been implemented; among them, an original procedure for converting multidimensional Fourier transforms and circular convolutions into one-dimensional and vice versa (Torras, 1983), another for the spatial and temporal aggregation of discharges, one of demarcation of regions in the network by similitude of their neurons' firing frequency and 3-D representation of these frequencies, together with the usual procedures to compute and represent histograms, bar diagrams and curves of various sorts.

In order to facilitate the interpretation of the graphs that will be included when analyzing the simulation results, we next describe the genesis and the information contained in each of them.

- **Temporal recordings of membrane potential and discharges:** these reflect the evolution of the variables P and y, respectively. They can be obtained at the level of the individual neurons (e.g., Figure 3.2), of a group of neurons (e.g., Figure 5.16(b)) or of the whole net (e.g., Figure 5.10(b) and (c)). The recordings of the latter type, which we will call "global", provide a first indication of the degree of synchronization with the stimulus and of the learning rate; those of the second type permit locating the region where the learning of a specific frequency predominates.

- **Discrete Fourier Transform:** its application to recordings of global membrane potential (e.g., Figure 5.10(d)) has a special interest, since it permits detecting increases or decreases in the amplitude of specific frequency components. For analyzing the impulse recordings of the neurons' outputs, it is instead preferable to use histograms of interspike intervals.

- **Curve of evolution of the synchronization with the stimulus:** it displays the network output aggregated in a time interval (usually of 20 ms) centered in the stimulation impulse, for several successive impulses (e.g., Figure 5.8). This curve allows one to observe recruiting phenomena (progressive synchronization) and frequency learning phenomena.

- **Histograms of interspike intervals:** these supply, for each neuron, information about the dominant period as well as about the dispersion around it (e.g., Figure 5.15(c)). From them, the average interspike intervals and standard deviations for all neurons are computed and graphically represented through a bar diagram.

- **Bar diagram:** the height of each bar is proportional to the number of neurons that have the corresponding average firing period, while the width is proportional to the average of the standard deviations associated with these neurons (e.g., Figure 5.5(b)).

- **3-D representation of the average interspike intervals of the neurons in the network:** it allows one to detect, by visual inspection, regions that have the same firing frequency (e.g., Figure 5.9(a)).

- **Demarcation of the regions where a specific frequency predominates** (e.g., Figure 5.9(b)): it is obtained by means of an image segmentation technique called "region growing" (Ernst et al., 1976; Pavlidis, 1982), consisting in adjoining to one or more initial points, supplied as starting data, those of their neighbors with a value no further from the initial one than a prefixed quantity and proceeding next in the same way with the adjoined points.

5.3 - SIMULATION RESULTS

> *"Si empezaba a tirar del ovillo iba a salir una hebra de lana, metros de lana, lanada, lanagnórisis, lanatúrner, lannapurna, lanatomía, lanata, lanatalidad, lanacionalidad, lanaturalidad, la lana hasta lanáusea pero nunca el ovillo".*
>
> *J. Cortázar, 1963*

Rather than exhaustively describing the results obtained in the simulations performed, we intend to offer in the present section a synthetic and interpretive view of the way certain factors influence the behavior of the network. The factors chosen are: the connectivity, the intraneuronal parameters, the initial state and the stimulation conditions. With this purpose, we will first describe the results of an experiment that will later be used as reference when analyzing the effect of the factors listed above.

5.3.1 - Reference experiment

The values of the parameters that define the connectivity of the network used in this experiment are n=3, p=1/3, a_e=1 and a_i=4. For the parameters that characterize the intraneuronal functioning, the assignment stated in Section 3.3 has been maintained and, in the two cases not specified there, the values $\sigma_{Pb_1}=0$ and c=0.1 have been initially fixed, to later explore in Section 5.3.2.2 the consequences derived from their variation.

The running in process consisted of 40,000 iterations; during the first 20,000, the index ξ, computed every 500 iterations corresponding to 5 ms each, has shown a decaying evolution from the initial value 4.0, to attain a stationary regimen around the value 1.4 in the last 20,000 iterations (Table 5.2).

Time	Index ξ	Time	Index ξ
5 s	4.044	105 s	1.729
	3.668		1.285
10 s	3.339	110 s	1.833
	3.733		1.543
15 s	3.409	115 s	1.652
	3.303		1.637
20 s	3.324	120 s	1.342
	3.371		1.619
25 s	3.089	125 s	1.540
	2.708		1.533
30 s	2.577	130 s	1.345
	2.706		1.817
35 s	2.880	135 s	1.584
	2.403		1.432
40 s	2.653	140 s	1.477
	1.900		1.721
45 s	2.198	145 s	1.225
	2.392		1.526
50 s	2.719	150 s	1.432
	1.993		1.453
55 s	2.045	155 s	1.527
	2.131		1.308
60 s	1.775	160 s	1.460
	2.067		1.439
65 s	2.209	165 s	1.594
	2.083		1.435
70 s	2.263	170 s	1.536
	2.133		1.428
75 s	1.807	175 s	1.506
	1.830		1.162
80 s	1.718	180 s	1.682
	2.207		1.453
85 s	1.806	185 s	1.323
	1.594		1.077
90 s	1.572	190 s	1.246
	1.559		1.334
95 s	1.655	195 s	1.546
	1.799		1.439
100 s	1.664	200 s	1.349
	1.327		1.533

*** Table 5.2 - Evolution of the index ξ, computed every 2.5 s, during the running in process.

We will next describe the results obtained in three successive situations of the learning process: before submitting the network to any stimulation frequency, while being presented with the first frequency and while being presented with the second frequency. For each situation we will state the network response to several rhythmic stimuli, previously presented or not, and the distribution of frequencies when the stimulation is stopped.

5.3.1.1 - Behavior before learning

Figure 5.5 shows the spatial distribution of average frequencies in the network when there is no stimulation, as well as the bar diagram that indicates not only the number of neurons that fire at each frequency, but also the dispersion of their interspike intervals. There seems to be a clear predominance of high frequencies (i.e. shorter periods) in the front part of graph (a) and of low frequencies in the back right part. This anisotropy is due to randomness in the initial distribution of frequencies in the network, and not to an asymmetry in the metric of the space. High frequencies show in general a greater dispersion than the low ones. For later comparative purposes, the Fourier transform of the spontaneous global membrane potential during a time interval of 12.8 s is also included (graph (c)).

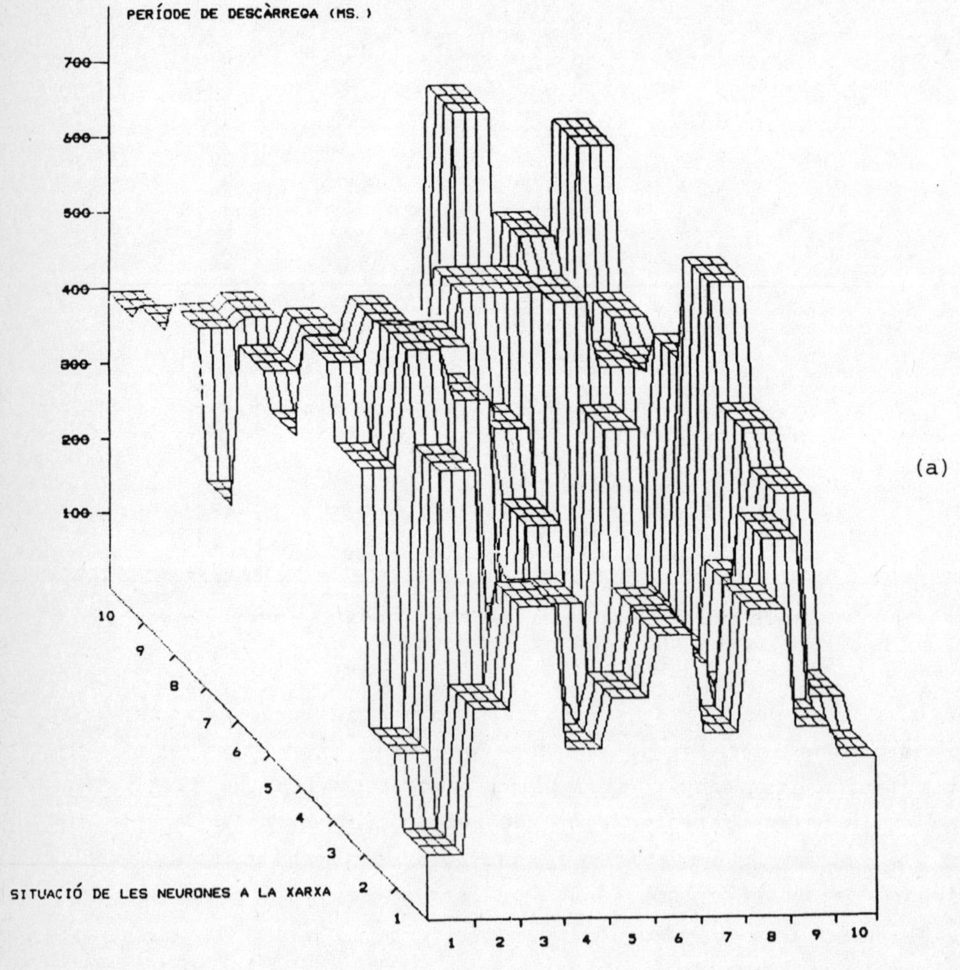

(a)

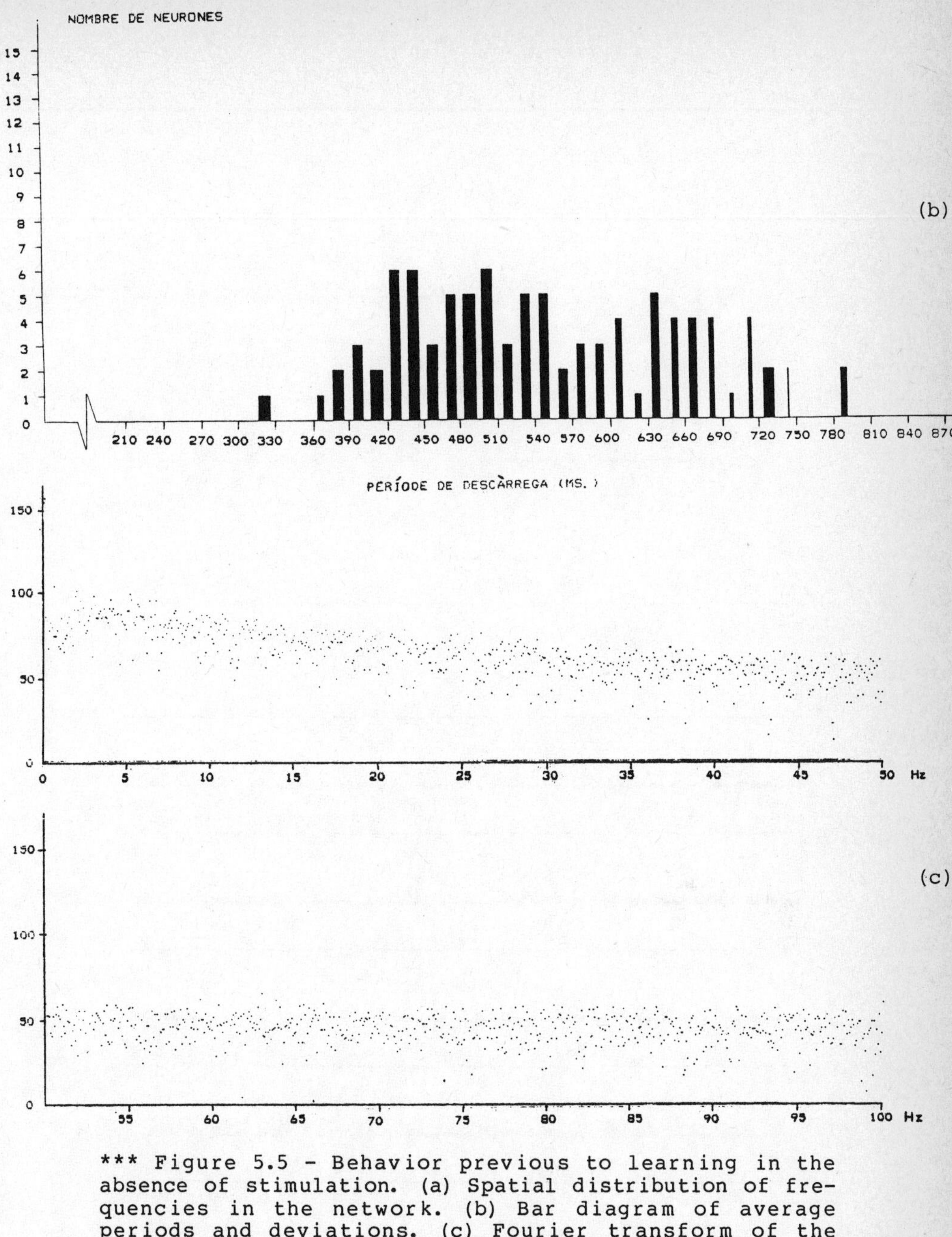

*** Figure 5.5 - Behavior previous to learning in the absence of stimulation. (a) Spatial distribution of frequencies in the network. (b) Bar diagram of average periods and deviations. (c) Fourier transform of the spontaneous global membrane potential, during a time interval of 12.8 s (logarithmic scale).

Setting the constant that governs the learning rate to zero (c=0) to nullify its effect, and presenting the network with a rhythmic stimulus of period 400 ms, the response characterized in Figure 5.6 is obtained.

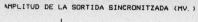

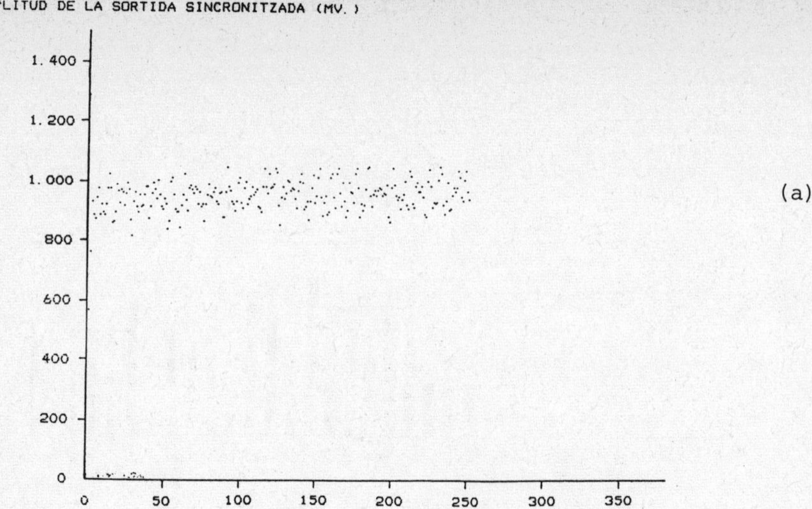
AMPLITUD DE LA SORTIDA SINCRONITZADA (MV.)
1.400
1.200
1.000
800
600
400
200
0
0
50
100
150
200
250
300
350
NOMBRE D'IMPULSOS D'ESTIMULACIÓ

(a)

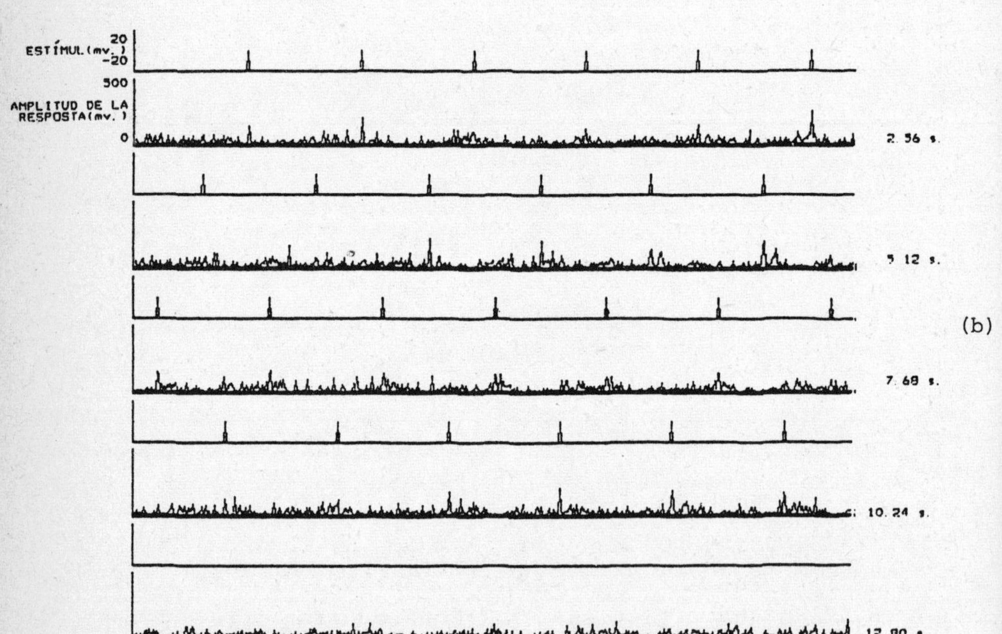
20
ESTÍMUL (mv.)
-20
500
AMPLITUD DE LA RESPOSTA (mv.)
0
2.56 s.
5.12 s.
7.68 s.
10.24 s.
12.80 s.

(b)

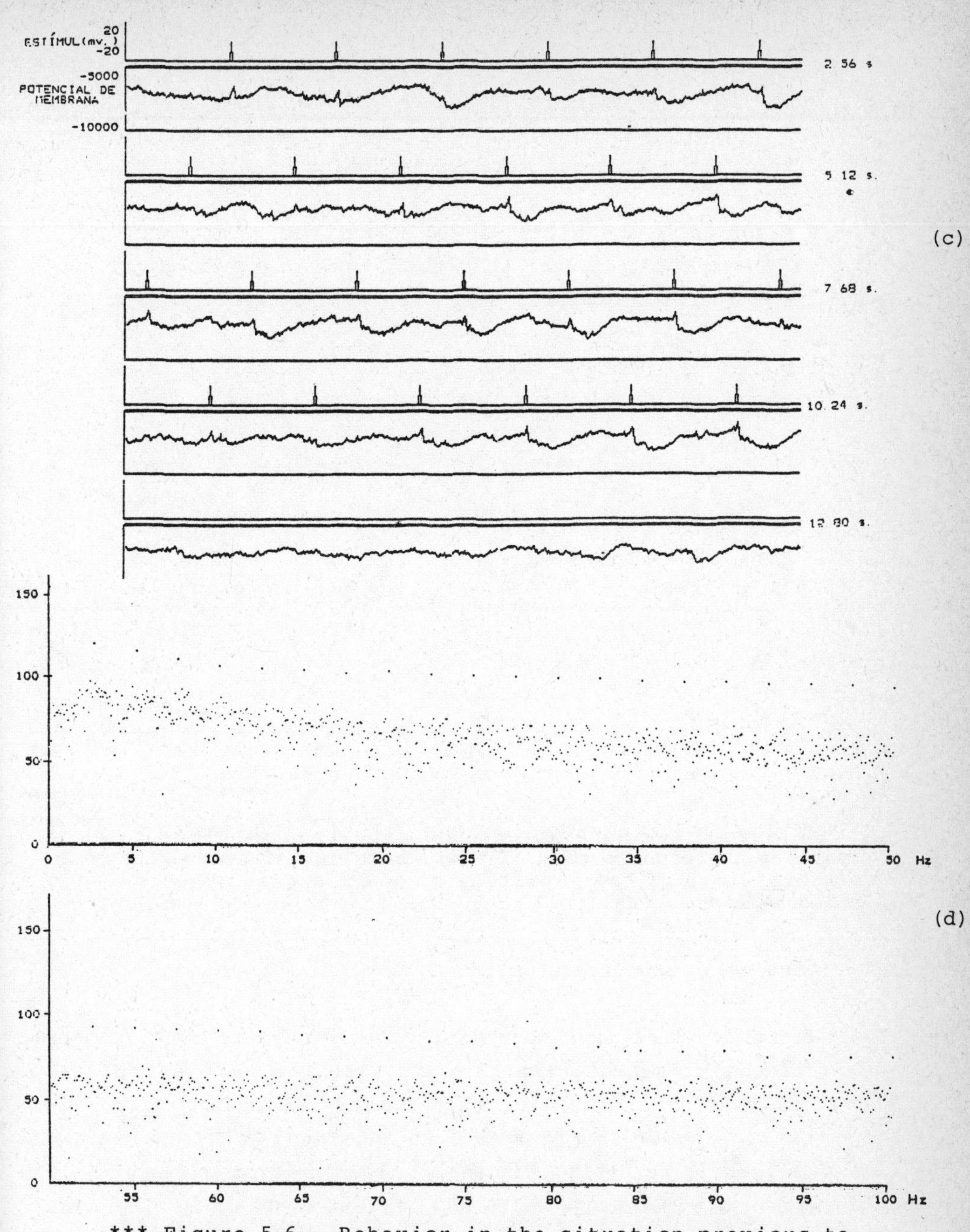

*** Figure 5.6 - Behavior in the situation previous to learning in response to a rhythmic stimulus of period 400 ms. (a) Evolution of the amplitude of the response synchronized with the stimulus. (b) Amplitude of the response during an interval of 12.8 s. (c) Amplitude of the membrane potential during the same interval. (d) Fourier transform of the membrane potential represented in the preceding graph (logarithmic scale).

The synchronization with the stimulus is low (graphs (b) and (c)) and its evolution, stationary (graph (a)). The Fourier transform of the global membrane potential (graph (d)) shows a clear predominance of the stimulation frequency over the remaining ones, revealed by the appearance of spikes in the spectrum for the harmonics of 2.5 Hz.

The network responds in a similar way to a rhythmic stimulus of period 550 ms, with the only difference that the degree of synchronization is lower and the dispersion of periods is greater (Figure 5.7).

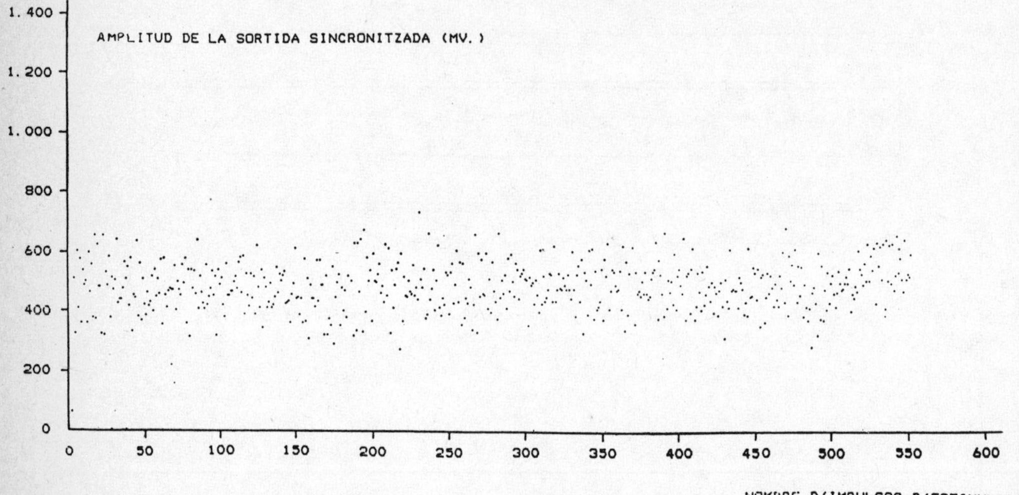

*** Figure 5.7 - Behavior in the situation previous to learning in response to a rhythmic stimulus of period 550 ms. Evolution of the amplitude of the response synchronized with the stimulus.

5.3.1.2 - Learning of one frequency

Setting now the constant that governs the learning rate to a positive value (c=0.1) and presenting the network with the same stimulus as before of period 400 ms, the evolution of the response synchronized with the stimulus shown in Figure 5.8 is obtained. After a transient regimen of slightly increasing tendency, a periodic steady regimen is attained, with maximum average value and minimum dispersion in the time interval represented (100 s). Table 5.3 contains a listing of the values taken by the index ξ, computed every 2.5 s.

From comparison of graphs 5.6(a) and 5.8, the following conclusion can be extracted: the learning process produces an increase in the amplitude of the response synchronized with the stimulus of approximately 25% and a decrease in its dispersion through time of approximately 45%.

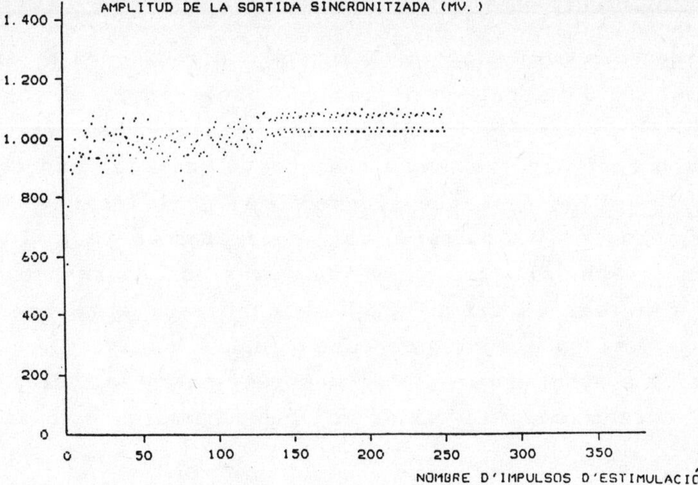

*** Figure 5.8 - Evolution of the amplitude of the response synchronized with a stimulus of period 400 ms, during the process of learning this stimulus. Note that the steady regimen attained is periodic, the big intracycle differences resulting from changes in the number of neurons discharging synchronously with the stimulus and the small ones, from interchanges in the concrete neurons firing; although, in the final analysis, both are due to harmonics.

Time	Index ξ	Time	Index ξ
	1.584		.361
5 s	2.037	55 s	.812
	1.868		.656
10 s	1.034	60 s	.616
	.878		.200
15 s	.472	65 s	.200
	.671		.100
20 s	.510	70 s	.100
	.400		.000
25 s	.574	75 s	.000
	.436		.100
30 s	.458	80 s	.000
	.539		.200
35 s	.949	85 s	.100
	.693		.000
40 s	.812	90 s	.100
	.686		.100
45 s	.911	95 s	.000
	.721		.000
50 s	.714	100 s	.000

*** Table 5.3 - Evolution of the index ξ, computed every 2.5 s, during the process of learning a stimulus of period 400 ms. Observe that ξ decreases progressively toward zero (stationarity), small fluctuations around this value being due to randomness in the neurons' interspike interval.

Making the network evolve, once the stimulation has stopped at the end of the learning process, the spatial distribution of average frequencies and their dispersion through time represented in Figure 5.9 results. The perception of a region at the front left part of graph (a), of homogeneous period close to 400 ms, is confirmed by the outcome of the automatic demarcation of regions shown in graph (b). From its comparison with Figure 5.5 it follows that learning has settled down in a predisposed zone, in the sense that there the starting frequencies were not too distant from the learned one, while frequencies in the rest of the network have suffered only small random variations; in the bar diagram (graph (c)), not only the number of neurons that fire on average at the learned frequency has increased, but its dispersion through time has radically decreased; and finally, in the Fourier transform of the spontaneous global membrane potential (graph (d)), the components corresponding to 2.5 Hz and its harmonics have increased in number.

If, after a period of spontaneous evolution, the network is submitted once more to the stimulus previously presented and learned, the response shown in Figure 5.10 is obtained. The stationary regimen that appears in graph (a) presents characteristics very similar (with regard to periodicity, average value and dispersion) to those of the regimen attained at the end of the learning process (Figure 5.8), though their detailed structures differ slightly, because of the different configuration of neurons' phases at the moment of turning on the stimulation. The index takes in this episode values between 0 and 0.2. Graphs (b) and (c) show the evolution of the amplitude of the global output and of the global membrane potential at the start and the stop of the stimulation; in the former, the usual recruiting phenomenon (progressive synchronization) is clearly observed and, in the latter, so is the prediction by the network of the moment of arrival of the next stimulation impulse, described in Section 2.1.2. The Fourier transform of the global membrane potential (graph (d)) presents here spikes, for the stimulation frequency and its harmonics, of greater amplitude than those previously pointed out in commenting on Figure 5.6. The comparison of this figure with Figure 5.10 makes evident the effects the learning process has had upon the response of the network to the learned stimulus: strengthening of its frequency component and of the synchronization with it.

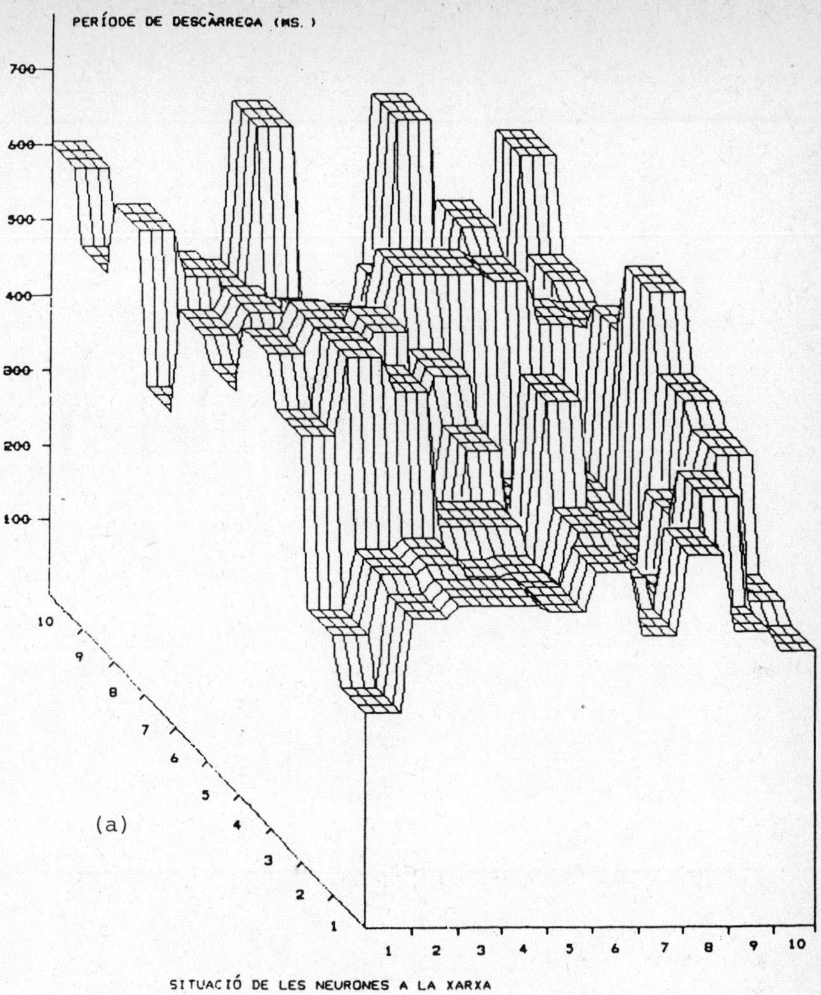

PERÍODE DE DESCÀRREGA (MS.)
700
600
500
400
300
200
100
10
9
8
7
6
5
4
3
2
1
1 2 3 4 5 6 7 8 9 10
SITUACIÓ DE LES NEURONES A LA XARXA

(a)

10
9
8
7
6
5
4
3
2
1
1 2 3 4 5 6 7 8 9 10

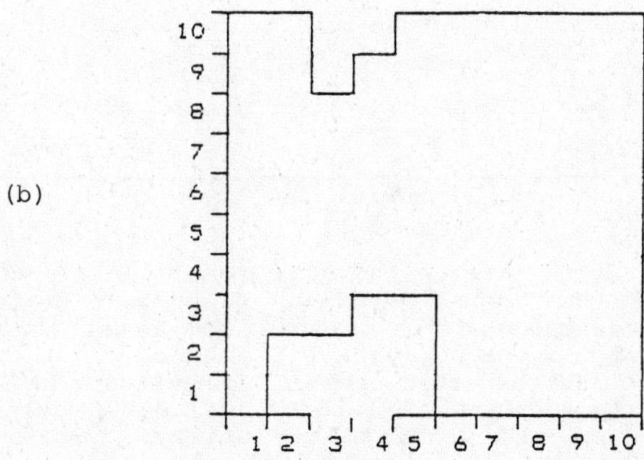

(b)

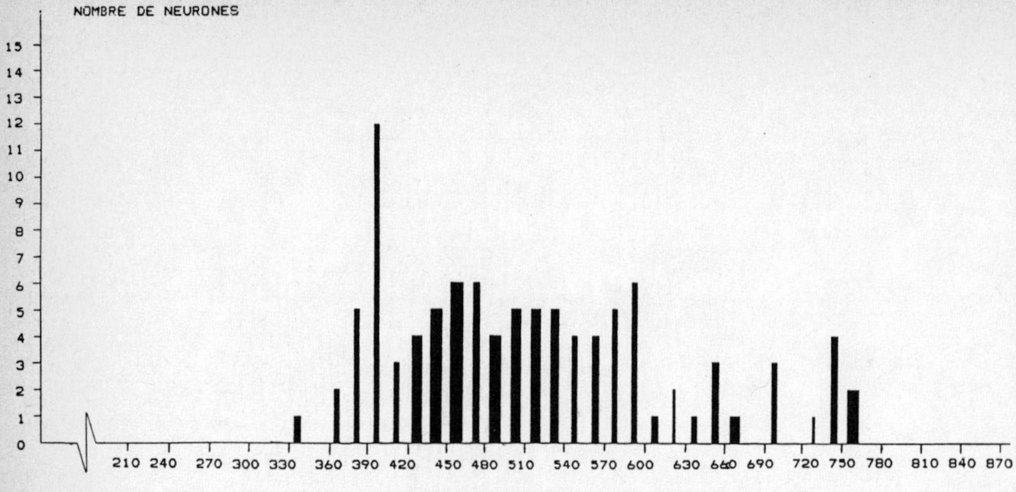

(c)

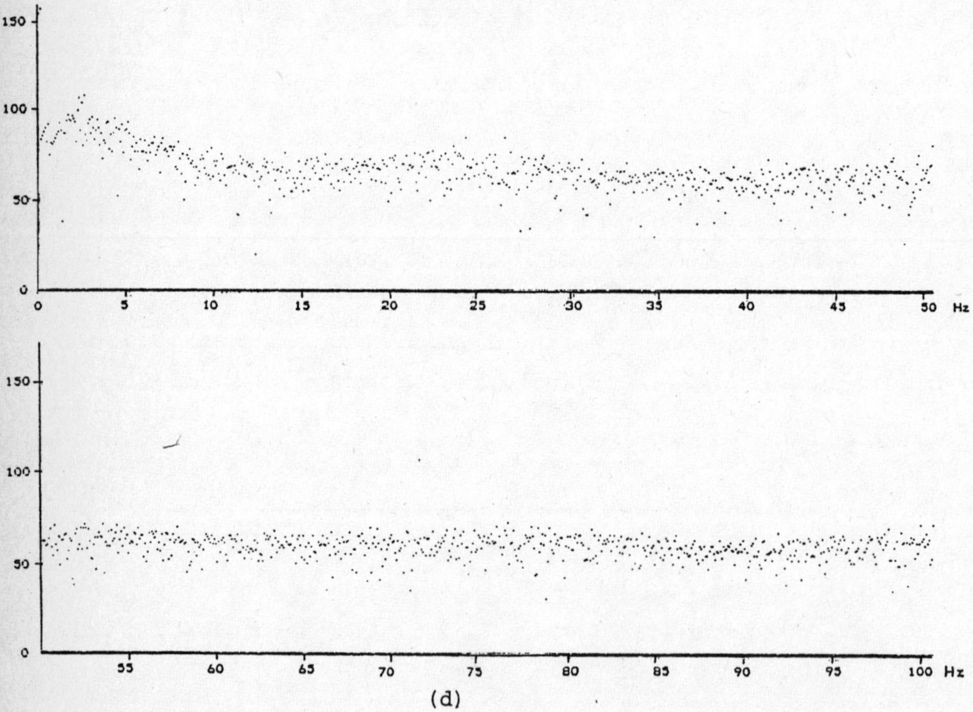

(d)

*** Figure 5.9 - State of the network at the end of the process of learning one frequency. (a) Spatial distribution of frequencies in the network. (b) Demarcation of the region of neurons of average frequency close to 2.5 Hz. The region is connected because connectivity has been defined on the surface of a torus. (c) Bar diagram of average periods and standard deviations. (d) Fourier transform of the spontaneous global membrane potential, during a time interval of 12.8 s (logarithmic scale).

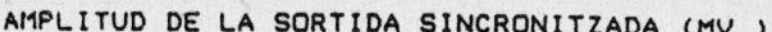
AMPLITUD DE LA SORTIDA SINCRONITZADA (MV.)

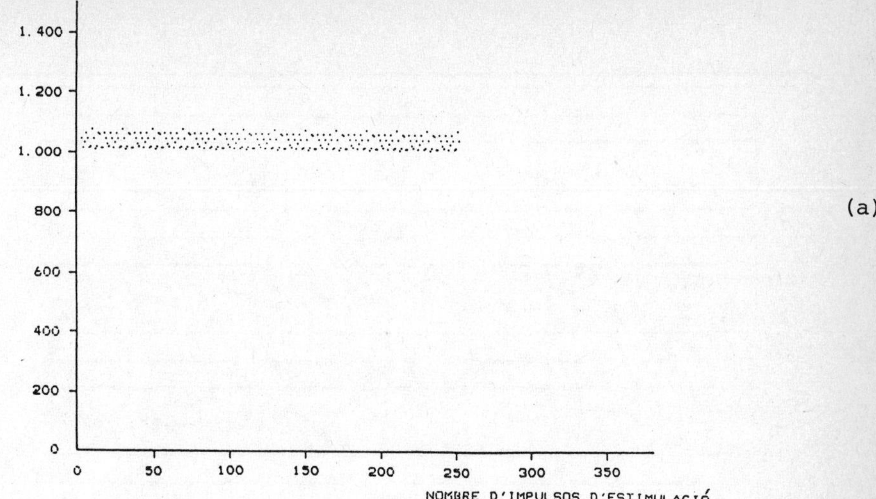
1.400
1.200
1.000
800
600
400
200
0
0
50
100
150
200
250
300
350
NOMBRE D'IMPULSOS D'ESTIMULACIÓ

(a)

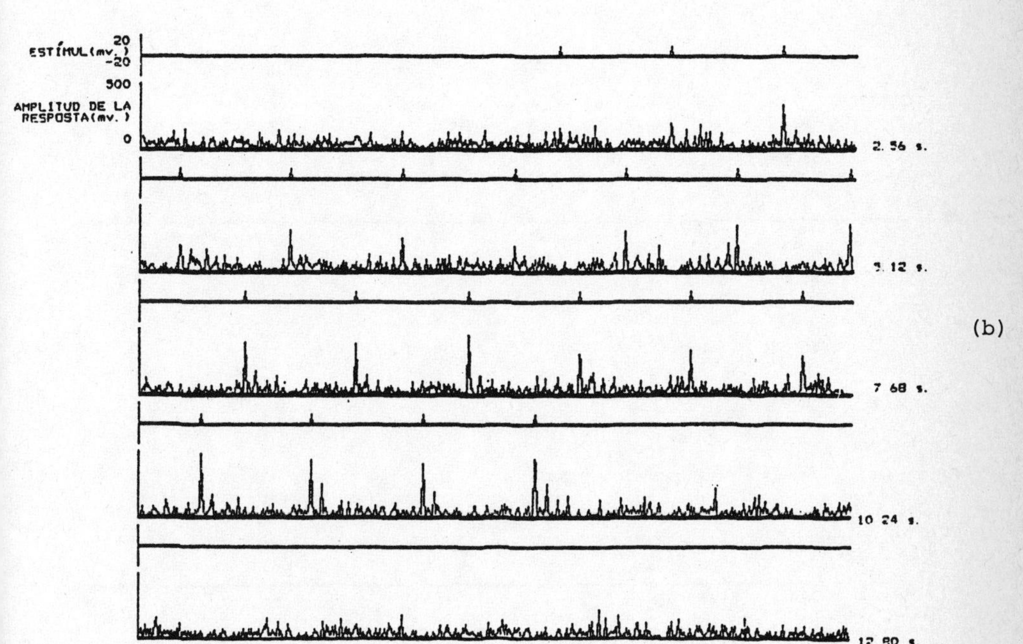
20
ESTÍMUL(mv.)
-20
500
AMPLITUD DE LA
RESPOSTA(mv.)
0
2.56 s.
5.12 s.
7.68 s.
10.24 s.
12.80 s.

(b)

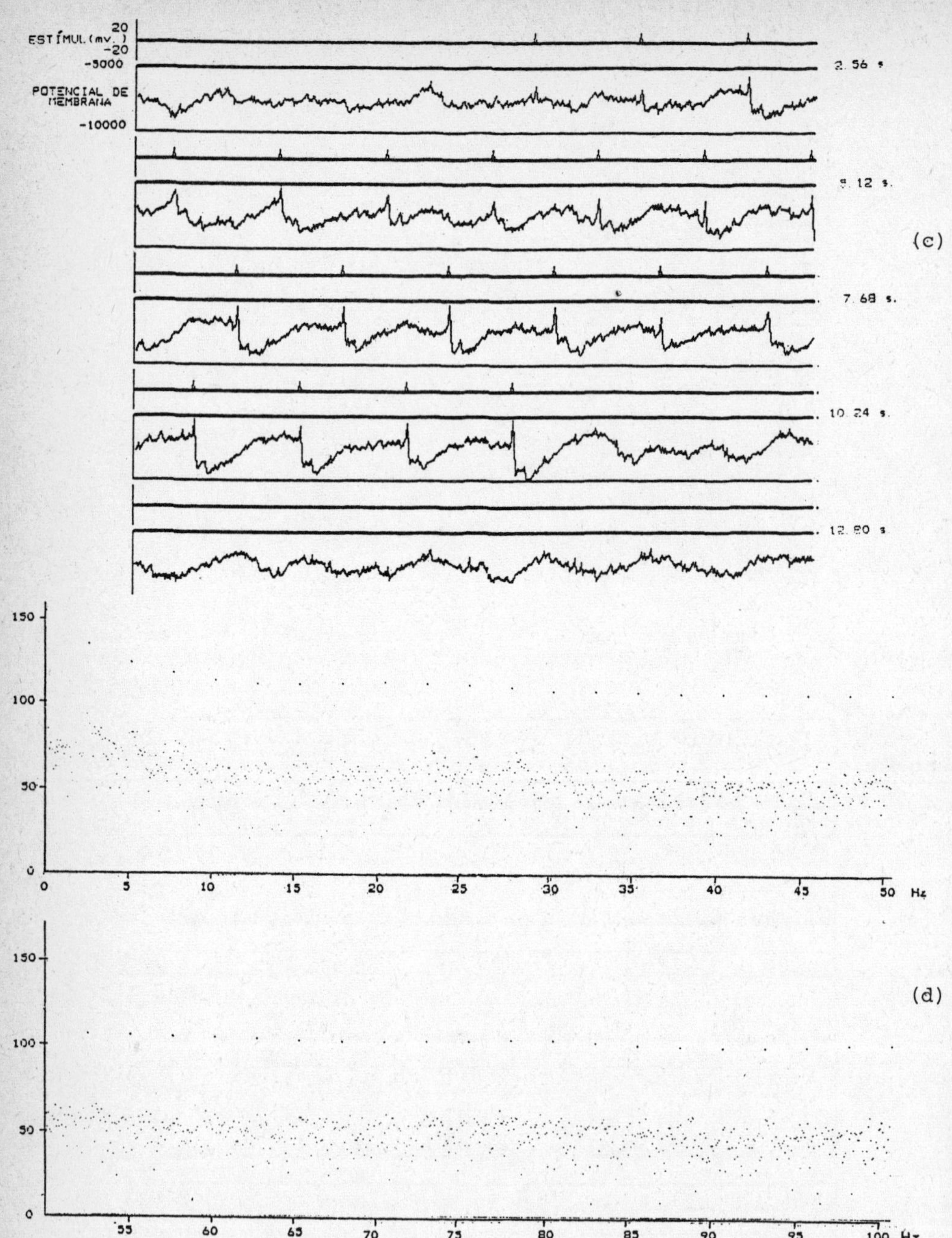

*** Figure 5.10 - Response to a previously-learned stimulus. (a) Evolution of the amplitude of the response synchronized with the stimulus. (b) Amplitude of the response during a time interval of 12.8 s. (c) Evolution of the global membrane potential during that same interval. (d) Fourier transform of the membrane potential represented in the preceding graph (logarithmic scale).

For stimuli of frequencies close to the learned one, a certain generalization of the effects of learning upon the response is observed: increase in the amplitude of the synchronized response, prediction of the time of arrival of the next stimulation impulse and coexistence of the frequency components of the presented stimulus and the learned one. Generalization is more radical for stimuli of higher frequency than the learned one, than for those of lower frequency than the learned one (Figures 5.11 and 5.12). Neither the confirmation nor the refutation at an experimental level of this fact has been reported.

The response of the network with c=0 to a rhythmic stimulus of period 550 ms (Figure 5.13) --quite different, therefore, from the learned one-- does not significantly diverge from the one previously shown in Figure 5.7, when the learning process had not yet taken place.

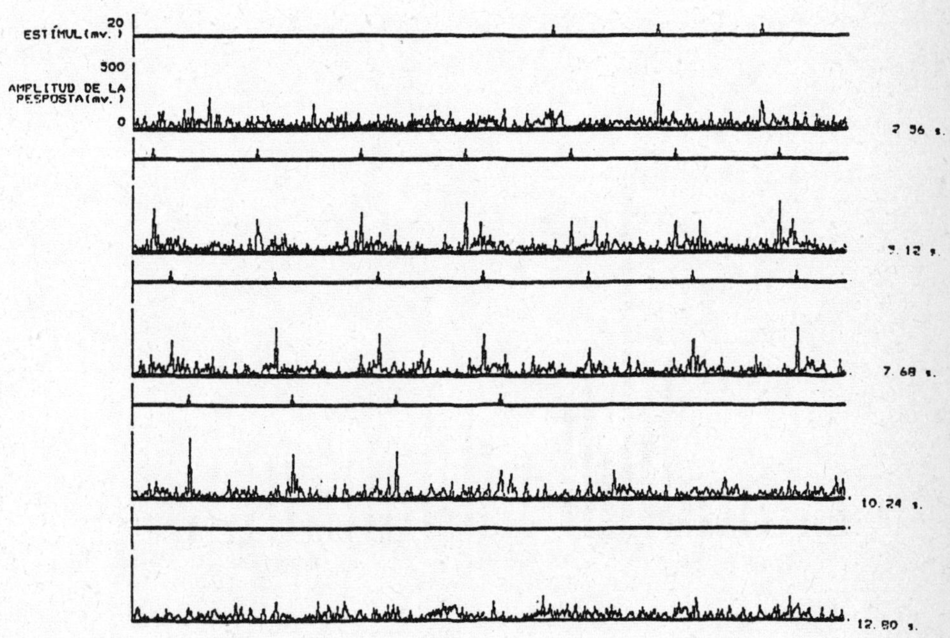

(a)

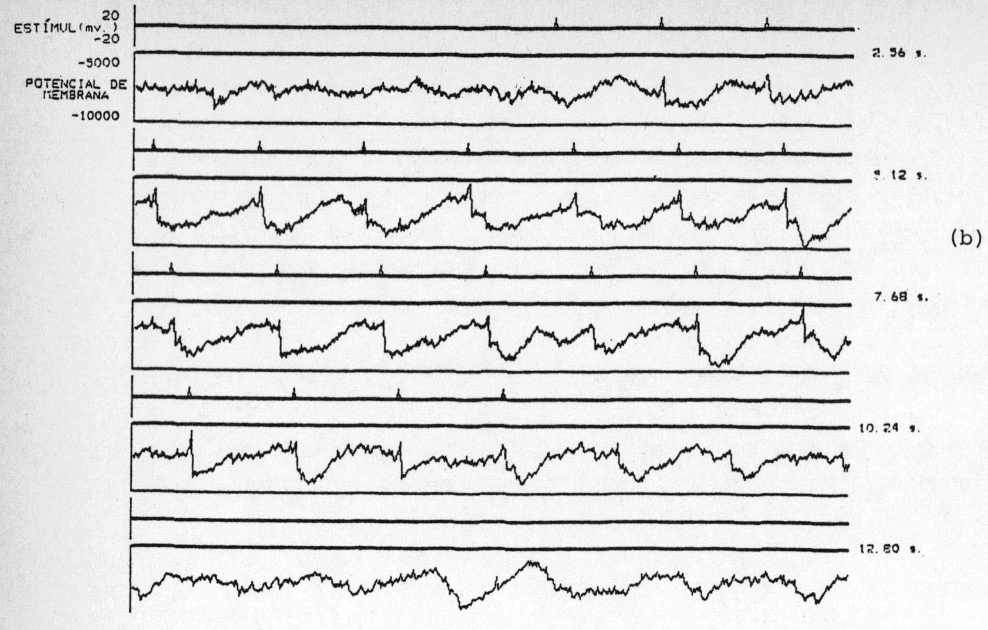

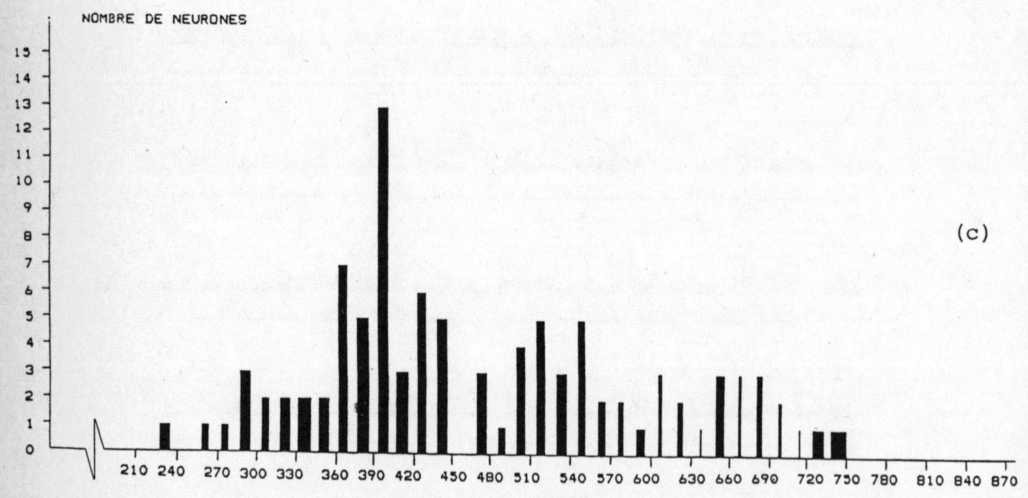

*** Figure 5.11 - Behavior subsequent to the learning of one frequency of 2.5 Hz: Response to a stimulus of period 385 ms. (a) Evolution of the output amplitude during a time interval of 12.8 s. (b) Evolution of the global membrane potential during that same interval. (c) Bar diagram of average periods and deviations.

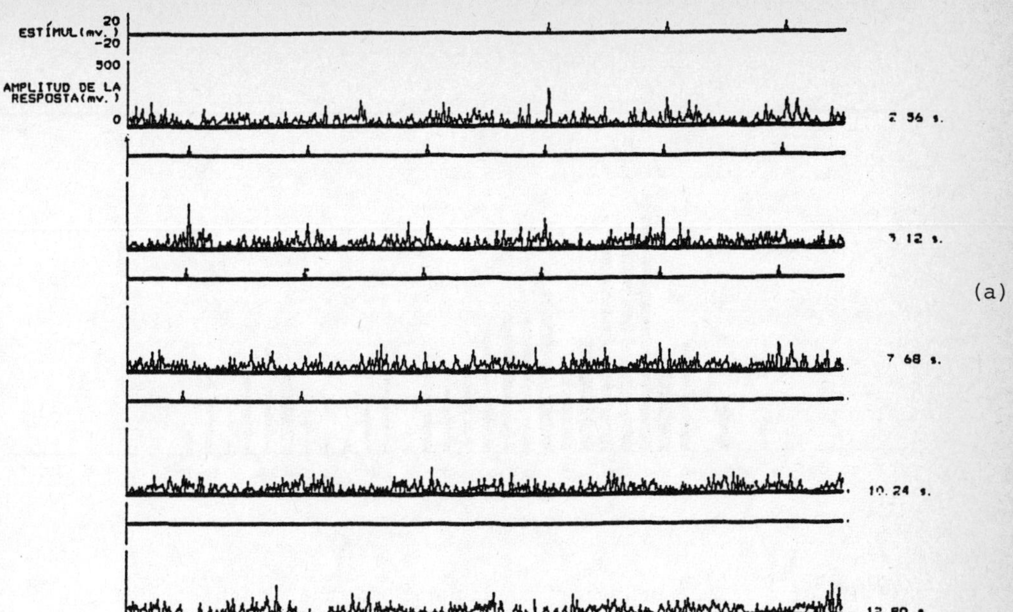
20
ESTÍMUL(mv.)
-20
500
AMPLITUD DE LA
RESPOSTA(mv.)
0
2.56 s.
5.12 s.
7.68 s.
10.24 s.
12.80 s.

(a)

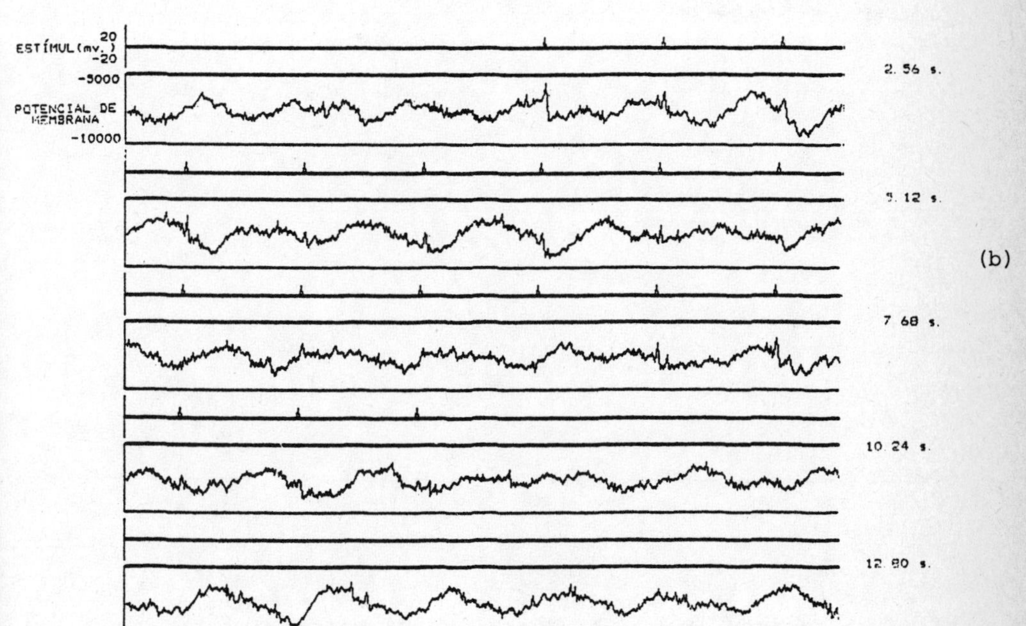
20
ESTÍMUL(mv.)
-20
-5000
POTENCIAL DE
MEMBRANA
-10000
2.56 s.
5.12 s.
7.68 s.
10.24 s.
12.80 s.

(b)

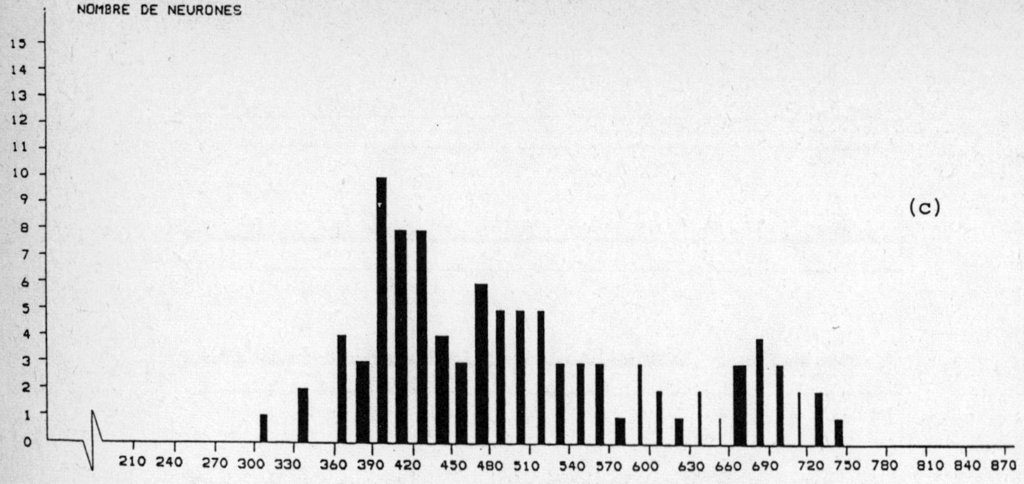

*** Figure 5.12 - Behavior subsequent to the learning of one frequency of 2.5 Hz: Response to a stimulus of period 420 ms. (a) Evolution of the output amplitude during a time interval of 12.8 s. (b) Evolution of the global membrane potential during that same interval. (c) Bar diagram of average periods and deviations.

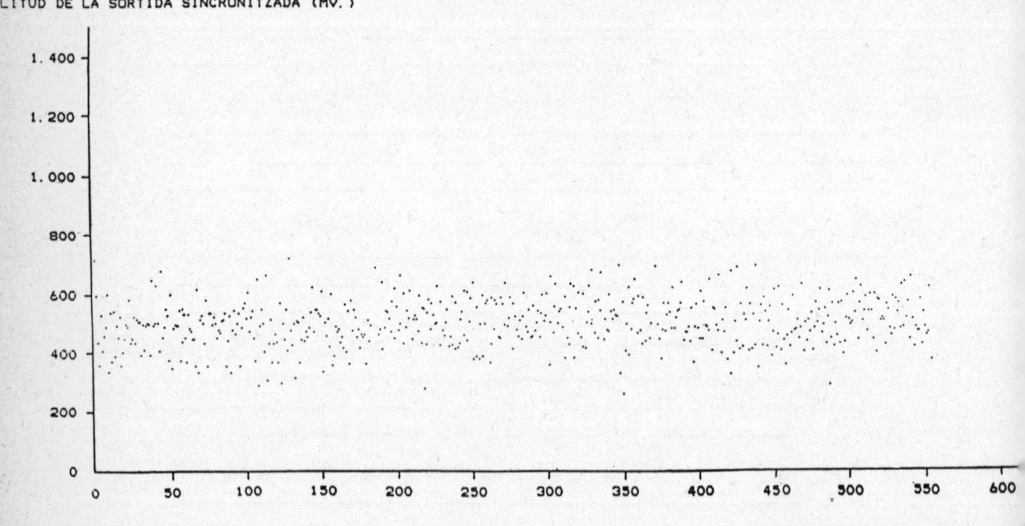

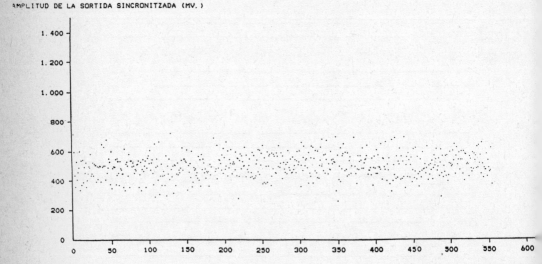

*** Figure 5.13 - Behavior subsequent to the learning of one frequency of 2.5 Hz: Evolution of the amplitude of the response synchronized with a stimulus of period 550 ms.

5.3.1.3 - Learning of two frequencies

Setting the constant that governs the learning rate back to 0.1 and submitting the network to a stimulus of period 550 ms, the evolution of the output synchronized with the stimulus shown in Figure 5.14 is obtained. The global trend of the distribution of points is clearly increasing with regard to the average value, and decreasing with regard to the dispersion. Both effects of the learning process are more obviously made evident by comparing this evolution of the synchronization with the stimulus, with those corresponding to the behavior previous to the learning of any frequency (Figure 5.7) and to the behavior subsequent to the learning of the first frequency (Figure 5.13): the increase in the average value is approximately 30% and the decrease in the dispersion, 50%. In the present case, a steady regimen neither as exactly periodic as the one shown in Figure 5.8 --though the final behavior is markedly oscillatory and stationary-- nor as clearly convergent --since the index ξ fluctuates between 0 and 0.6-- is attained. Furthermore, the evolution is slower and the steady regimen does not appear until after 2 minutes (24,000 iterations) of the learning process.

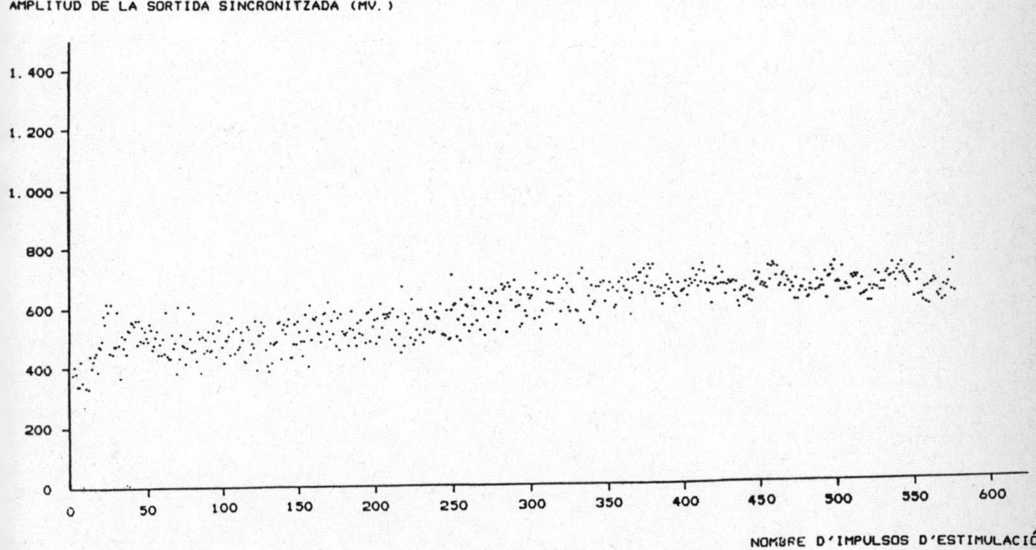

*** Figure 5.14 - Process of learning the second frequency: Evolution of the amplitude of the response synchronized with a stimulus of period 550 ms.

The spatial distribution of frequencies in the network and their dispersion through time resulting from the learning process are represented in Figure 5.15. Two regions of rather uniform frequency are discerned: one located in the left front part, of relatively short period, and the other established behind it toward the center of the network, of longer period. Graph (b) shows the demarcation of these regions --which we will denote A and B-- whose average frequencies are actually those of the two learned stimuli. The region of period 400 ms coincides, except for two neurons, with that resulting from the process of learning the first frequency (Figure 5.9). The region of period 550 ms has been also established in a predisposed area, in the sense previously expressed that the starting frequencies were already relatively low (Figure 5.5). In graph (c), the histograms of interspike intervals corresponding to the neurons in the fifth and sixth columns of the network are represented. The neurons in regions A and B show unimodal histograms, 400 ms being the mode of those corresponding to the first region and 550 ms the mode of those correspondig to the second region. Between both regions and behind the second one, there are some neurons with arhythmic behavior and diverse average periods. The bar diagram (graph (d)) permits appreciating in a more compact form the predominance of the two learned frequencies, as well as their lower dispersion in general than that of the remaining frequencies.

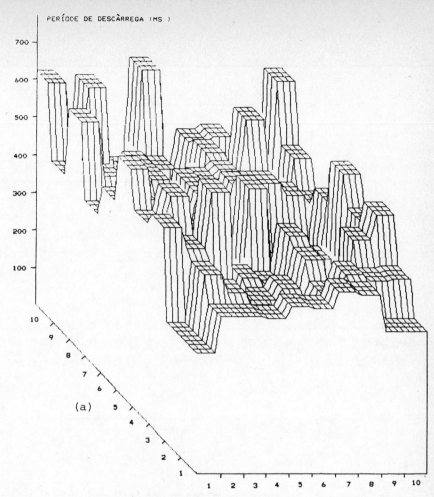

PERÍODE DE DESCÀRREGA (MS.)

700 600 500 400 300 200 100

10 9 8 7 6 5 4 3 2 1

1 2 3 4 5 6 7 8 9 10

SITUACIÓ DE LES NEURONES A LA XARXA

(a)

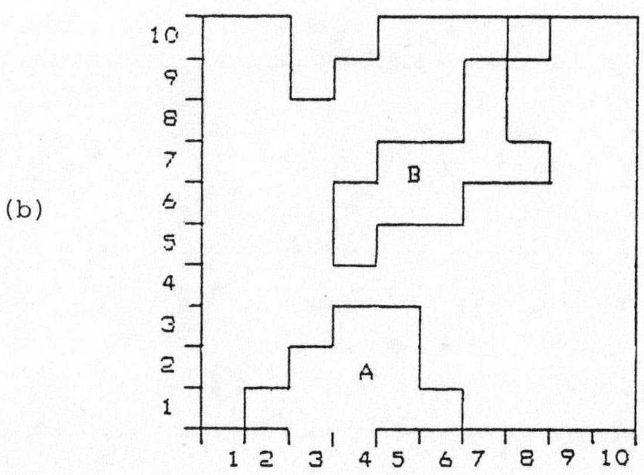

(b)

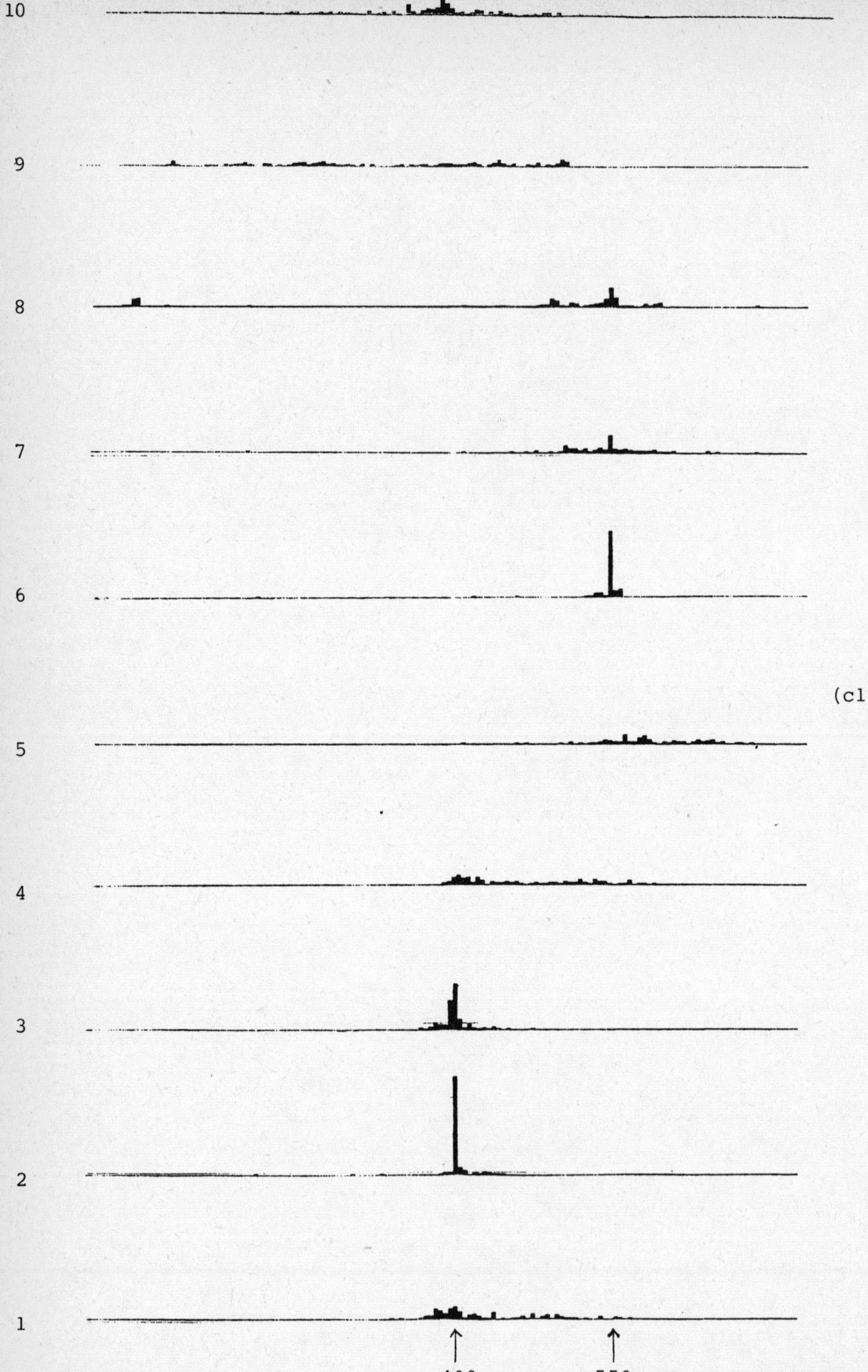
10
9
8
7
6
(c1)
5
4
3
2
1
400
550 ms.

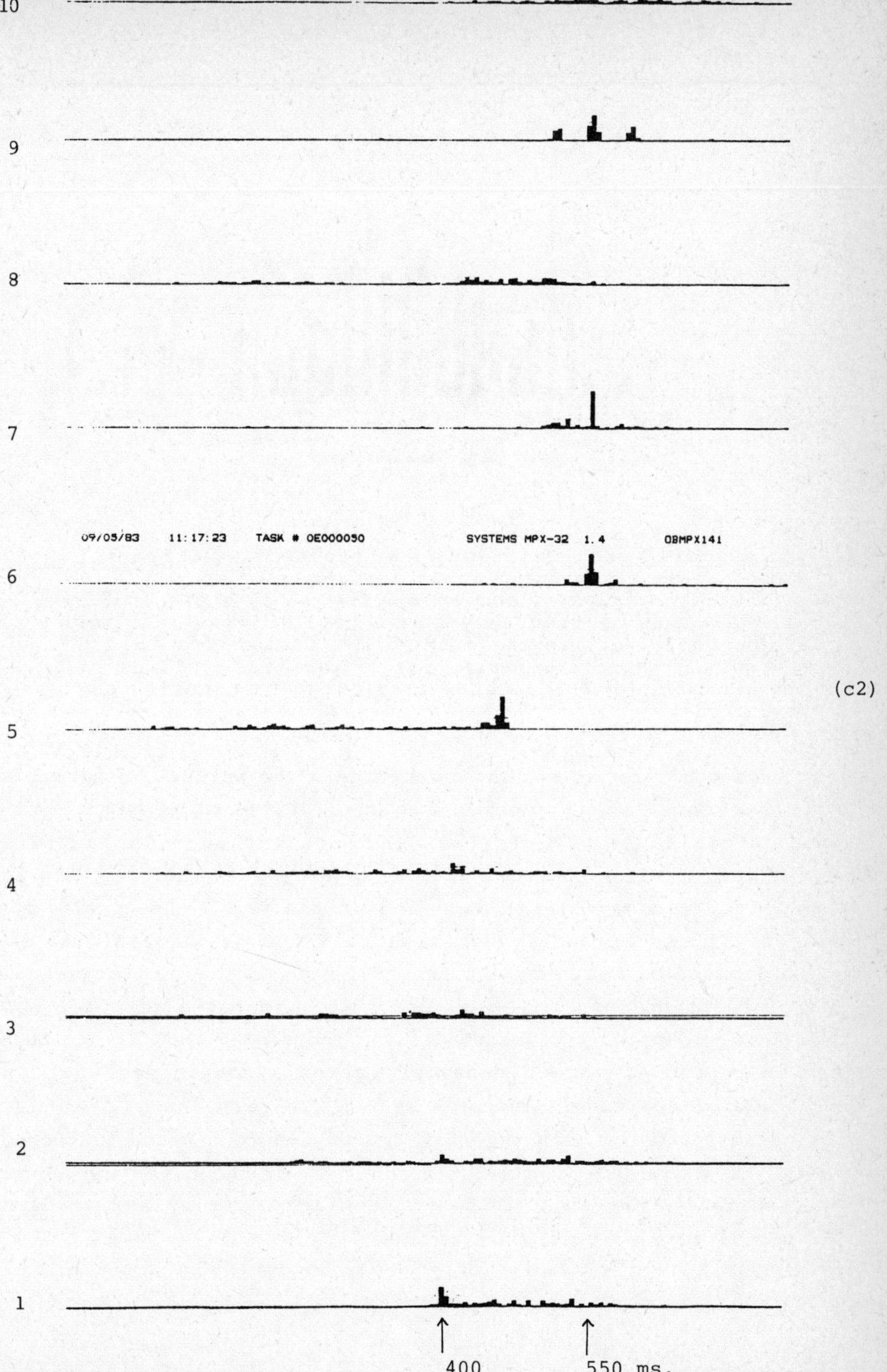
10
9
8
7
09/05/83 11:17:23 TASK # 0E000050 SYSTEMS MPX-32 1.4 OBMPX141
6
5
4
3
2
1
400
550 ms.

(c2)

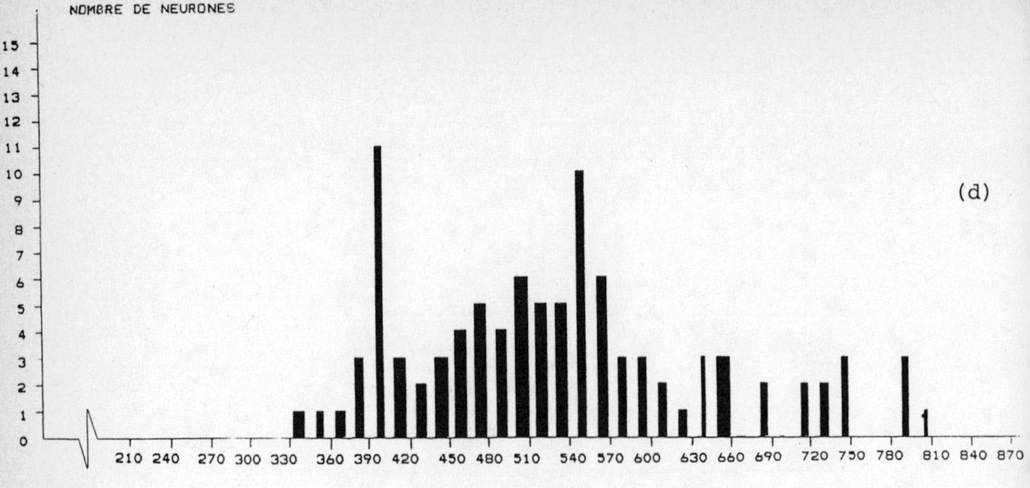

*** Figure 5.15 - State of the network at the end of the process of learning the second frequency. (a) Spatial distribution of frequencies in the network. (b) Demarcation of the groups of neurons of frequency close to 2.5 Hz --A-- and of frequency close to 1.81 Hz --B--. (c) Histograms of interspike intervals corresponding to the neurons in the fifth --c1-- and sixth --c2-- columns of the network. (d) Bar diagram of average frequencies and deviations.

If once the second frequency has been learned, the network is submitted again to the first of them, the response shown in Figure 5.16 is obtained. After a very brief (150 ms) transient regimen, a steady regimen of very similar characteristics (regarding periodicity, average value and dispersion) to that obtained at the end of the process of learning the first frequency is attained (Figures 5.8 and 5.10(a)). Thus, the process of learning the second frequency has not erased the effects of learning the first one. Graph (b) shows the detailed response of five groups of four neurons each. The second and fifth groups, located in region A, show a behavior rather synchronized with the stimulus and of the same frequency as it; the response of the third group is also periodic, although only two of its four neurons synchronize with the stimulus; finally, the behavior of the first and fourth groups, located in region B, is desynchronized and displays waves of greater period than that of the stimulus.

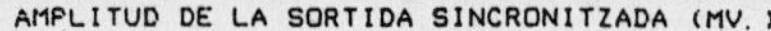

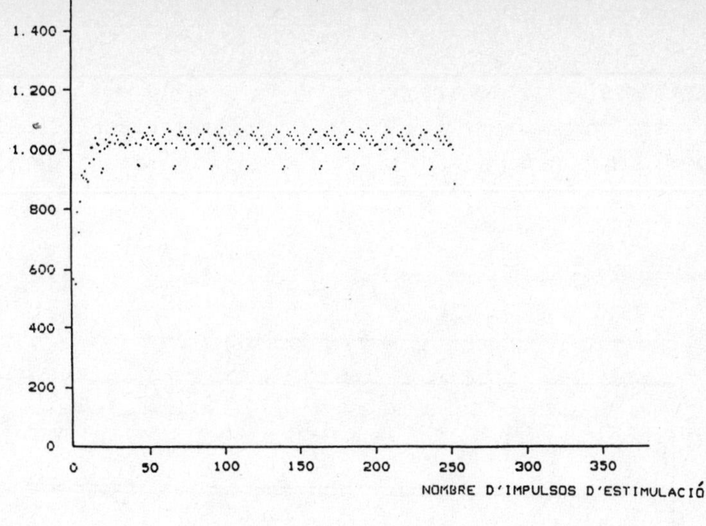

(a)

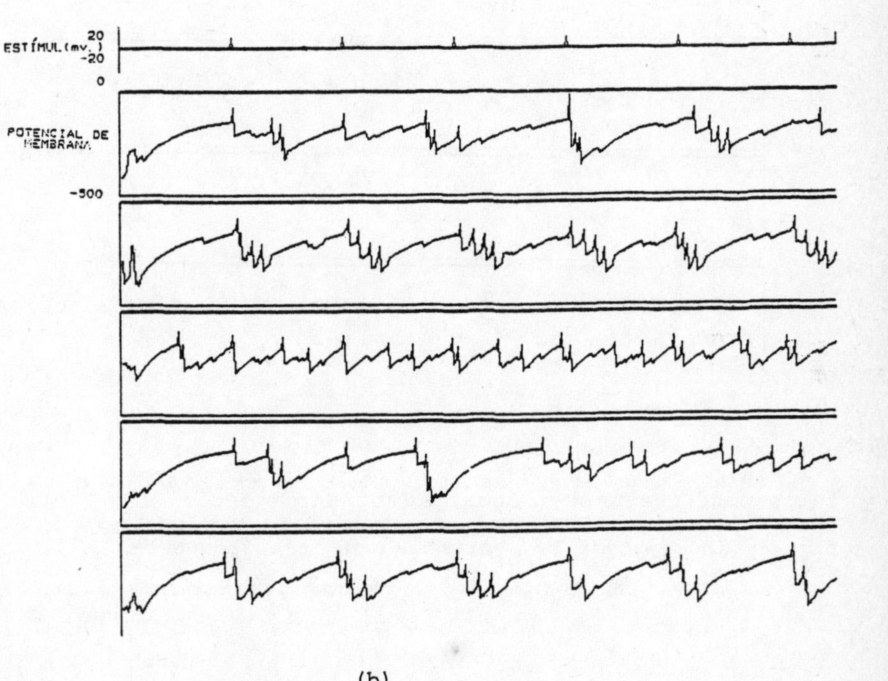

(b)

*** Figure 5.16 - Behavior subsequent to the learning of the second frequency: Response to a stimulus at the frequency learned in the first place (2.5 Hz). (a) Evolution of the amplitude of the response synchronized with the stimulus. (b) Local membrane potential for five groups of four neurons each.

Figure 5.17 shows the response of the same groups of neurons as above to rhythmic stimulation of period 550 ms. As expected, the first and fourth groups, which in the preceding recordings showed waves of period longer than 400 ms, synchronize now with the stimulus; the second group continues firing in phase at the frequency of the first stimulus, while the third and fifth groups show a rather arhythmic and desynchronized behavior.

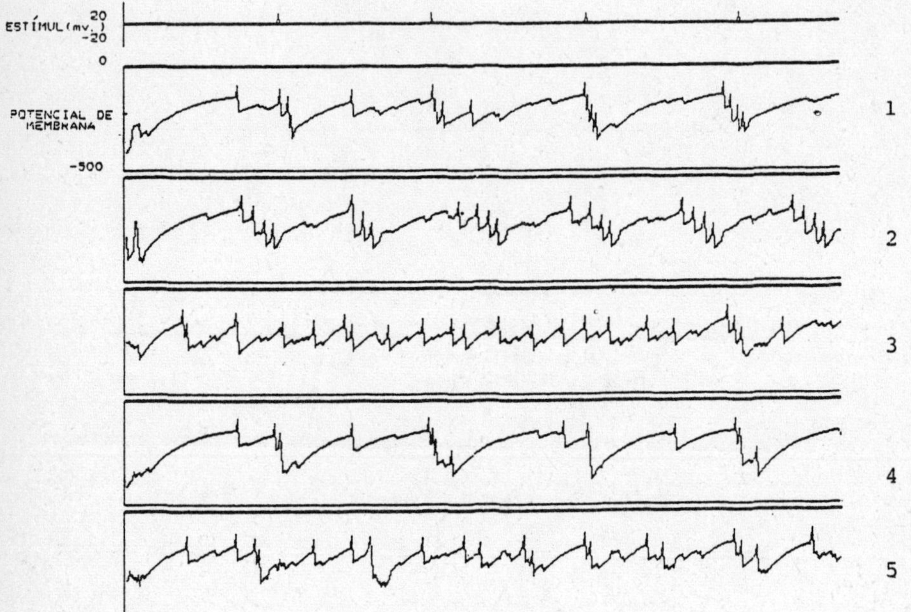

*** Figure 5.17 - Behavior subsequent to the learning of the second frequency: Local membrane potential for the same five groups of four neurons as in the preceding figure, in response to a stimulus of that frequency (1.81 Hz).

When presenting to the network a frequency intermediate between the two learned ones, it is observed that differentiated groups of neurons fire at each of the three learned frequencies. If the frequency presented is close to one of the learned ones, the predominance of this one upon the other is clear (Figure 5.18(a), (b), (c) and (d)); if, on the contrary, the period of the stimulus presented is equidistant from the two learned ones, no clear predominance of any of the three frequencies upon the other two is observed (Figure 5.18(e) and (f)).

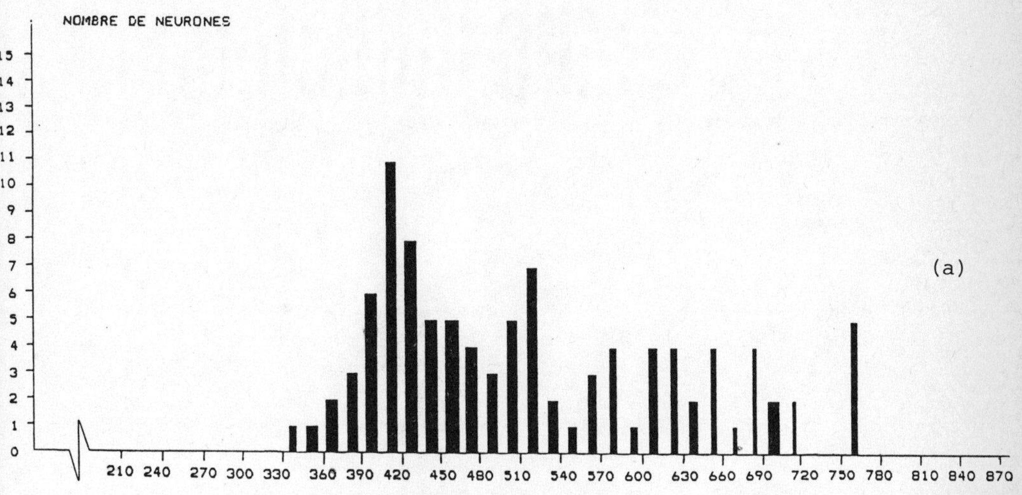

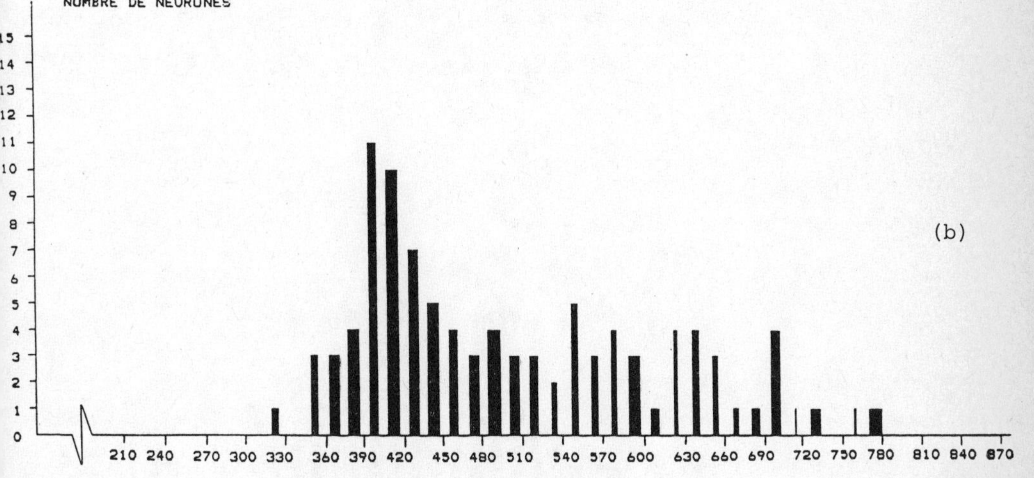

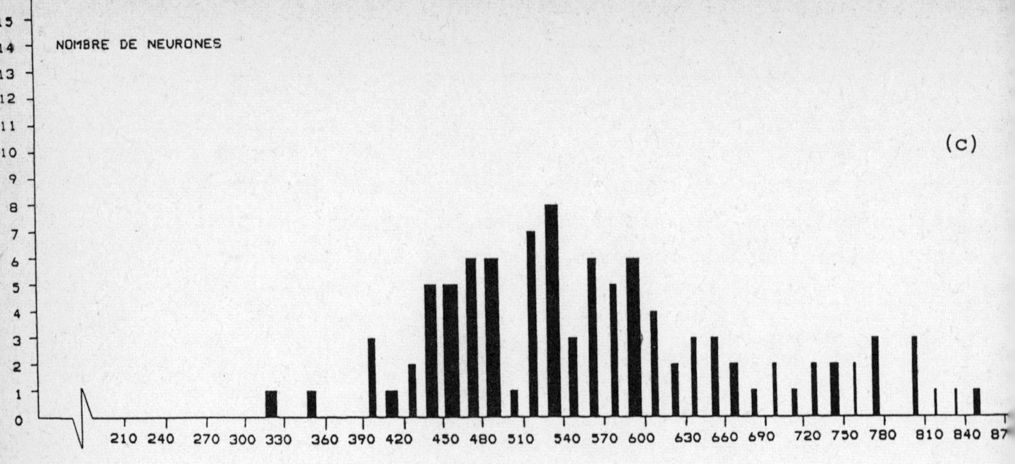
NOMBRE DE NEURONES
(c)
PERÍODE DE DESCÀRREGA (MS.)

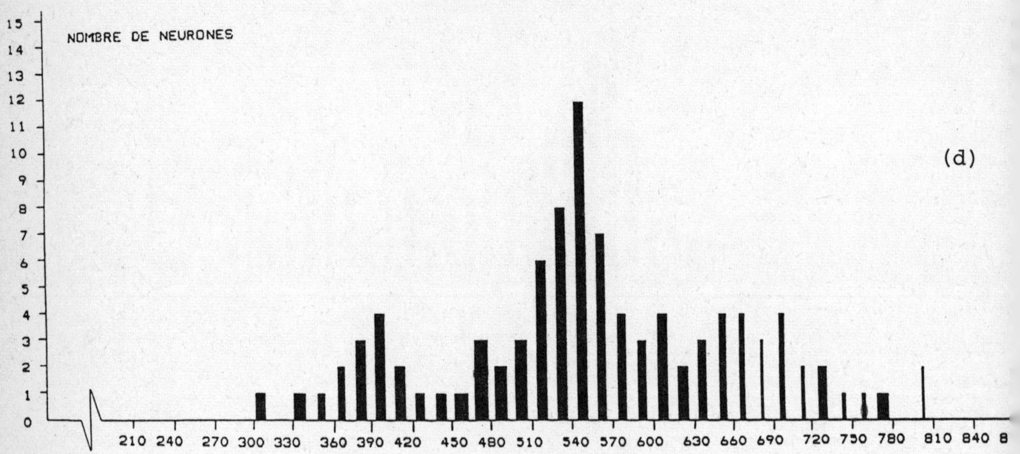
NOMBRE DE NEURONES
(d)
PERÍODE DE DESCÀRREGA (MS.)

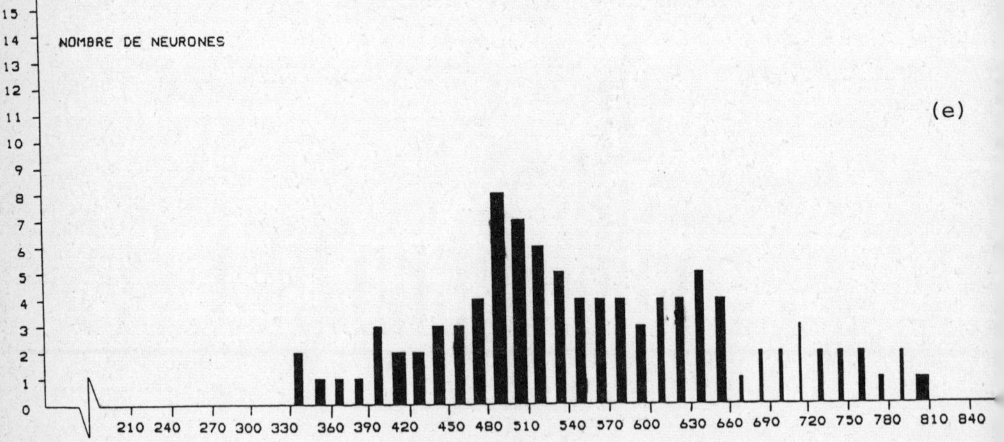
NOMBRE DE NEURONES
(e)
PERÍODE DE DESCÀRREGA (MS.)

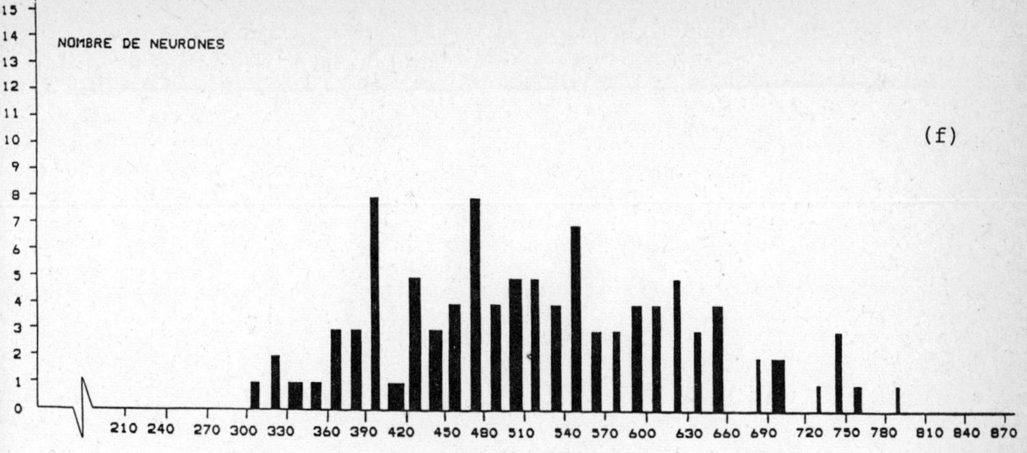

*** Figure 5.18 - Response of the network, before --(a), (c) and (e)-- and after --(b), (d) and (f)-- the learning process, to stimuli of periods intermediate between the two learned ones of 400 and 550 ms: (a) and (b) 415 ms; (c) and (d) 535 ms; (e) and (f) 480 ms.

5.3.2 - Effects of the variation of several factors

In the present section, we will analyze the implications of the variation of several parameters characterizing the connectivity, the intraneuronal functioning, the initial state and the stimulation conditions, upon the behavior of the network. We mention that, as a general rule, only the consequences of varying just one factor will be explored, maintaining the others equal to those used in the reference experiment. Not all the results obtained will be exhaustively included, but only the graphical representation of those that best illustrate the general trends observed; which after being detailed for each factor throughout the exposition, will be summarized in Table 5.3, included at the end of this section.

5.3.2.1 - Connectivity

We will consider five aspects: the number of synapses per neuron (n), the proportion of excitatory connections over the total (p), the ranges of the excitatory connectivity (a_e) and of the inhibitory one (a_i) arising from one neuron, the randomness in the microconnectivity and the overlap between excitation and inhibition.

The effect of the **number of synapses per neuron** has been explored for values of **n** between 0 and 10. Figure 5.19 shows, for n=0, the curves of evolution of the synchronization with the stimulus during the process of learning the two frequencies used in the reference experiment. The 34 neurons of initial m_{Pb_1} between -35.5 mv and -40.5 mv learn to fire with a period of 400 ms; while the 22 neurons of initial m_{Pb_1} between -42 mv and -45 mv learn to fire with a period of 550 ms. Both learnings take place here in a completely independent way, since they affect disjoint sets of non-interconnected neurons.

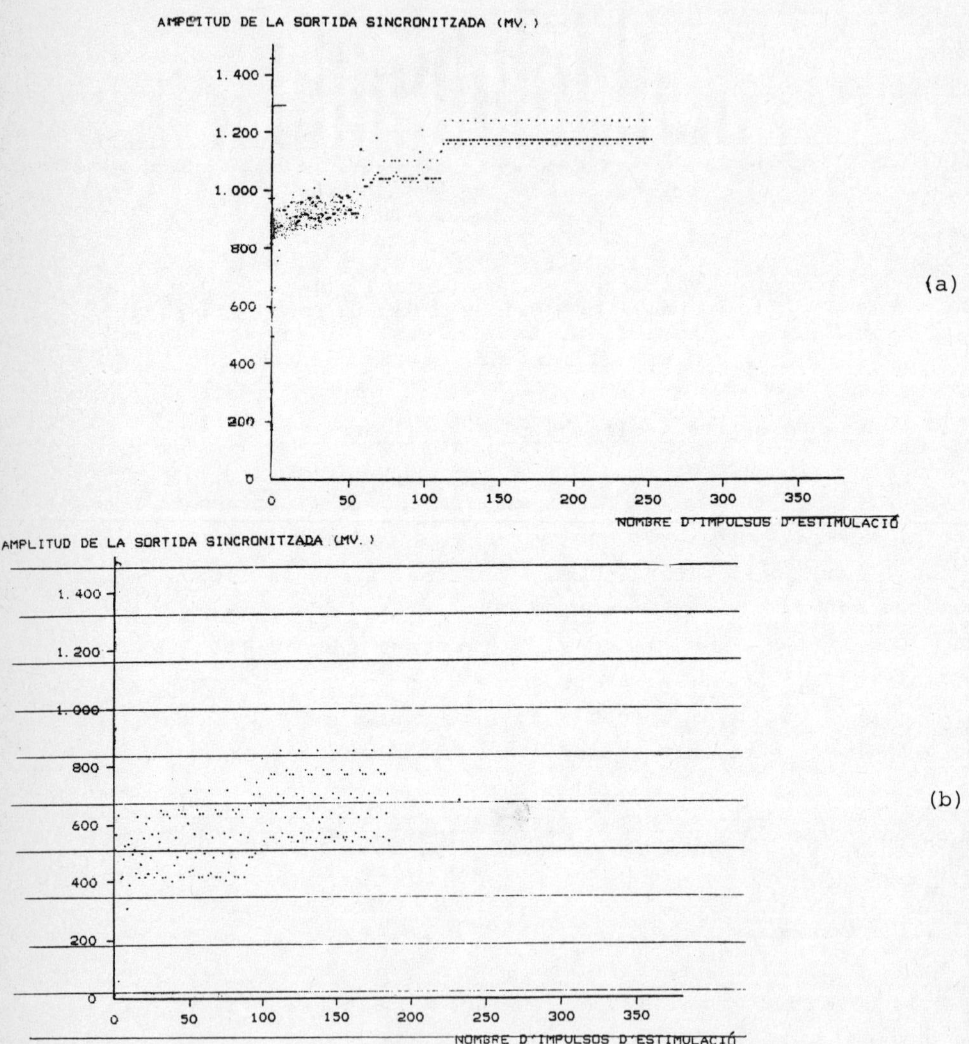

*** Figure 5.19 - Effect of the number of synapses per neuron (n): Evolution of the response synchronized with a stimulus of period 40 ms --(a)-- and with a stimulus of 550 ms --(b)--, for a set of non-interconnected neurons (n=0).

Comparing these graphs with those in Figures 5.8 and 5.14, and others of the same type corresponding to different values of n, it is observed that the increase in the number of connections lengthens the transient regimen and tends to diminish the dispersion through time of the amplitude of the synchronized response, hardly affecting its average level.

Concerning the **proportion of excitation**, the extreme cases p=0 and p=1 have been studied. In the former, the level of activity is considerably reduced with respect to the reference experiment, both in the spontaneous behavior and in response to stimulation; many neurons remain silent and, with the few exceptions of those deprived of incoming connections from other neurons because of randomness, the neurons that fire do so with a period greater than 650 ms. The running in process has absolutely no effect, since plasticity requires the presence of postsynaptic potentials to manifest itself. The only stimuli susceptible to being learned are those of period greater than 630 ms, though in this case learning does not take place in a region of the network, but only in isolated neurons.

In the opposite case (p=1), the level of activity is very high and the distribution of frequencies in the network is concentrated in a thin band at the upper extreme of the spectrum. The running in process does not show a tendency toward synchronization, at least during the initial 100,000 iterations (500 s), and learning can take place for only one frequency in that band, with which almost the entire network synchronizes in a brief time interval.

Networks with different **ratios between the ranges of the excitatory and the inhibitory connectivity** have been generated. In general, the greater a_e and the less a_i, the more concentrated the bar diagram that characterizes the distribution of frequencies in the network after the running in process (Figure 5.20) and, consequently, the more difficult the learning of more than one frequency. The average level of the amplitude of the response synchronized with a stimulus of frequency close to the average spontaneous one is considerably greater than that obtained in the reference experiment. In those networks where $2a_e+a_i>$ >10, the ranges of the incoming excitatory and inhibitory connectivity for each neuron overlap, making difficult the establishment of regions of uniform frequency, up to the point of precluding it in the extreme cases, in which learning takes place only in isolated neurons.

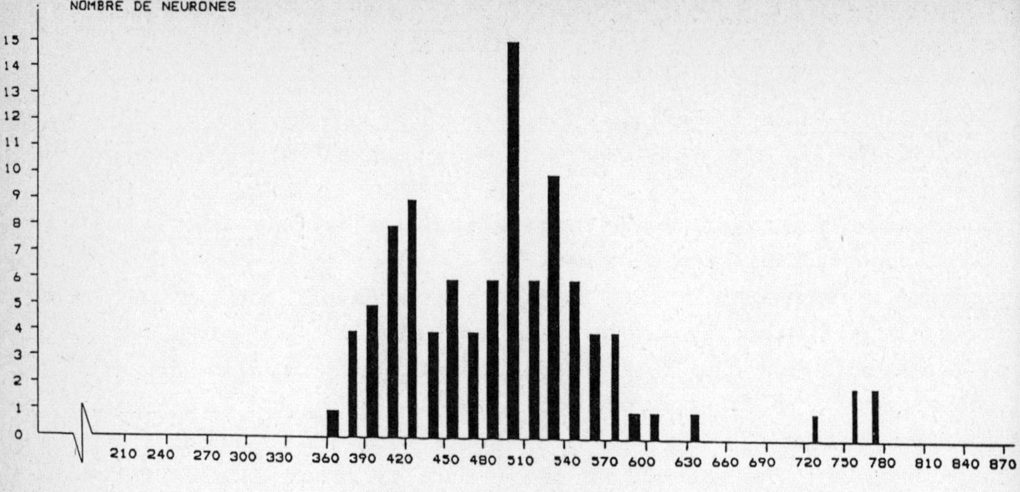

*** Figure 5.20 - Effect of the ratio between the ranges of the excitatory and the inhibitory connectivity (a_e/a_i): Bar diagram of average periods and deviations at the end of the running in process, for a network with a_e=4 and a_i=2.

Two networks with identical parameters, differing only in the **micro-connectivity** as a consequence of their random generation, behave in a very similar way in the several experimental situations considered, except with regard to the placement in the network of the regions where each of the learned frequencies predominates (Figure 5.21). In general, as we have previously noted, the mentioned frequencies tend to settle down in predisposed zones, whose location depends obviously on the microconnectivity. The fact that the behavior of the network is not sensitive to differences between several different realizations of the random connectivity justifies having defined the network model by means of a probability distribution and not through a fixed connectivity pattern.

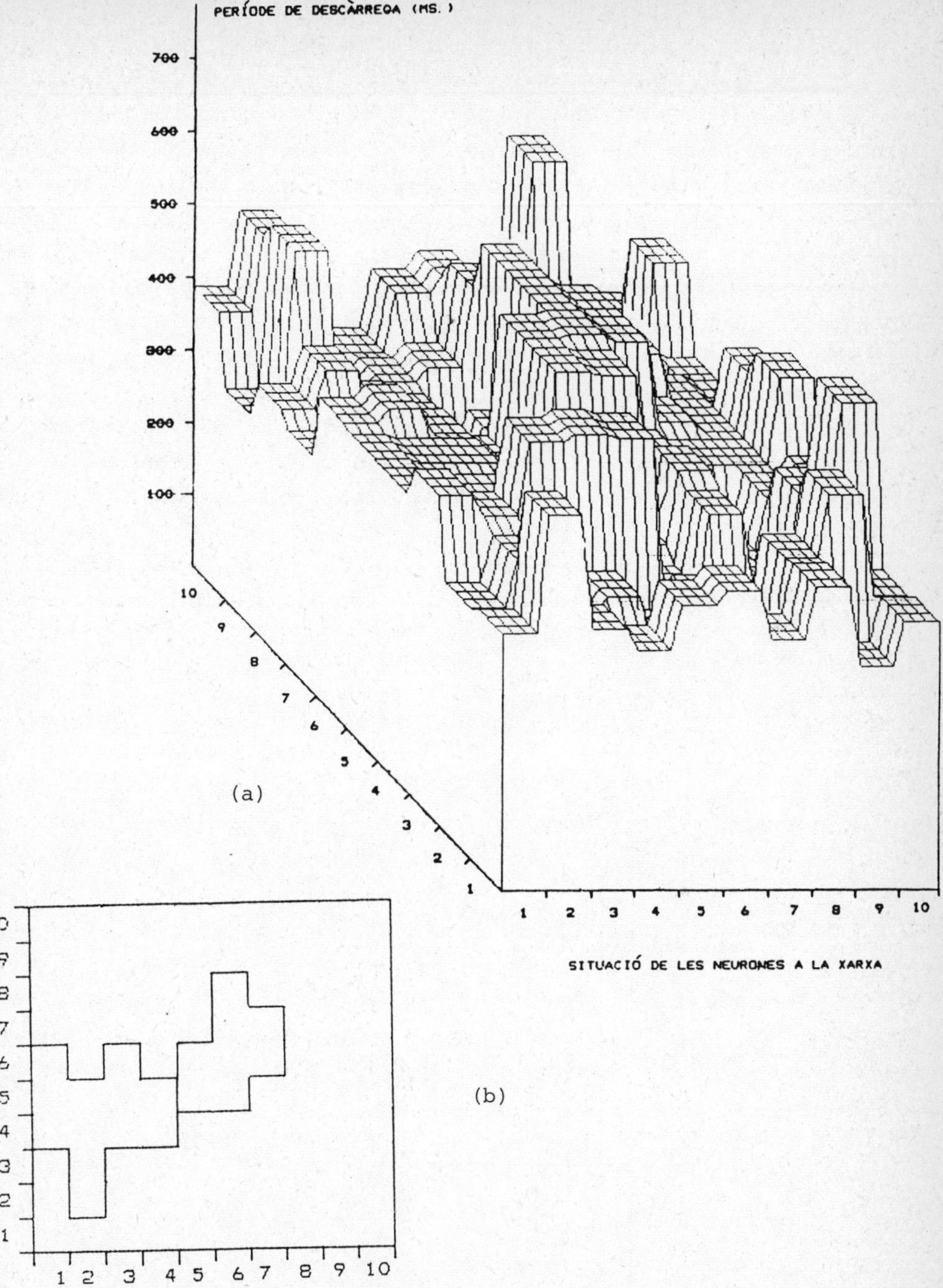

*** Figure 5.21 - Effect of microconnectivity. (a) Spatial distribution of frequencies in the network and (b) demarcation of the regions where each of the learned frequencies has settled down, for a network generated with the same parameter values than that of the reference experiment.

Finally, we have analyzed the effect that the change of the **probability density functions that govern the establishment of excitatory and inhibitory connections** has upon the behavior of the network. To be specific, we have generated networks using the functions shown in Figure 5.22(a), which entail the inclusion of the range of the excitatory connectivity within that of the inhibitory one, leading, therefore, to a lot of overlapping between them. The results obtained with these networks diverge from the ones obtained in the reference experiment basically in two aspects: the greater dispersion of the bar diagram that characterizes the distribution of frequencies in the network, both before and after learning, and the occurrence of learning in isolated neurons rather than in regions (Figure 5.22(b) and (c)). The last-mentioned fact supports the interpretation that the very shapes of the original probability density functions (Figure 5.2) are responsible for the learning of each frequency taking place in a region and not in isolated neurons, which is logical, taking into account that the synchronization between pacemaker neurons only excitatorily interconnected is more probable than when inhibitory connections are involved.

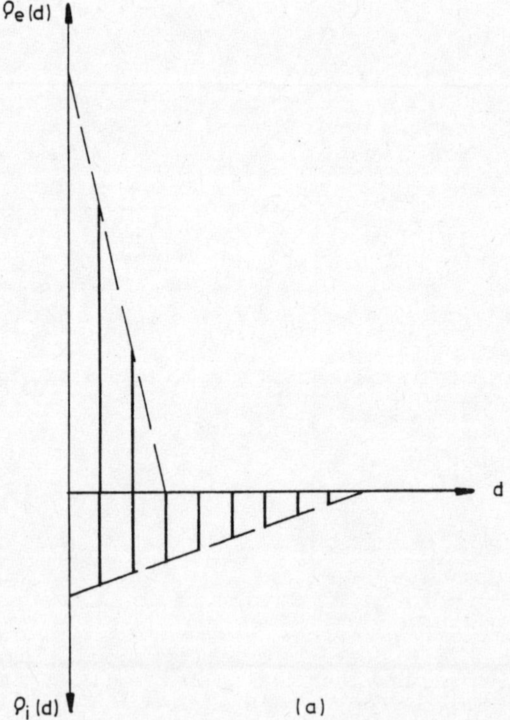

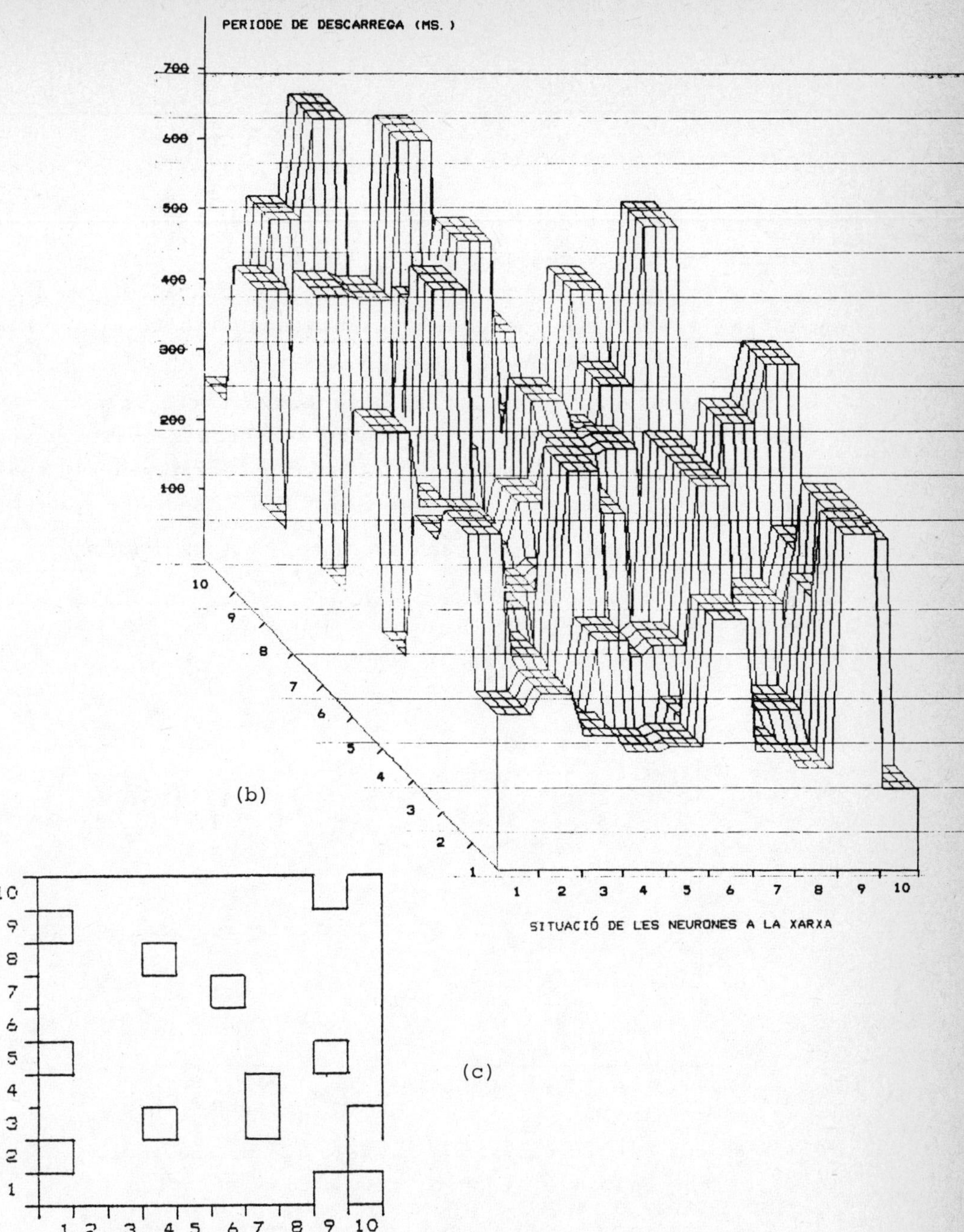

*** Figure 5.22 - Effect of the probability density functions that govern the establishment of excitatory and inhibitory connections. (a) Probability density functions used in the comparison. (b) Spatial distribution of frequencies in the network after the process of learning one frequency of 2.5 Hz. (c) Demarcation of the set of neurons which have learned this frequency.

5.3.2.2 - Intraneuronal parameters

We have explored the consequences of varying two intraneuronal parameters: the standard deviation of the asymptotic limit of the membrane potential (σ_{Pb_1}) and the learning rate (c).

Increasing the **standard deviation of Pb_1** leads to a delay in the attainment of the steady regimen during the process of learning one frequency, as well as to a loss of the periodic character of this regimen and a comparative increase of its dispersion through time (Figure 5.23). Above the value σ_{Pb_1}=3.5 mv, the effects of learning cannot be discerned, since the differences between the initial and final average values of the amplitude of the synchronized output and its dispersion stop being significant. It should be pointed out that the variation of σ_{Pb_1} does not affect the location in the network of the regions where the learning of each frequency settles down predominantly.

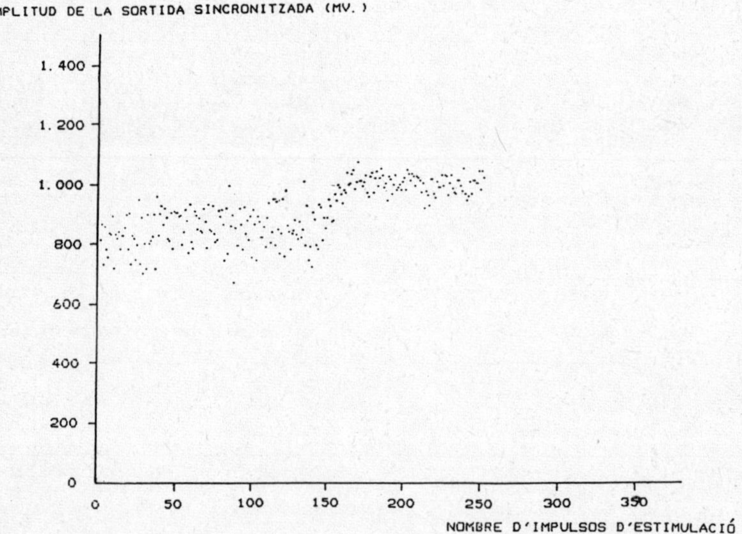

*** Figure 5.23 - Effect of σ_{Pb_1}: Evolution of the amplitude of the response synchronized with a stimulus of period 400 ms, during the process of learning this stimulus, for the same network used in the reference experiment, but with σ_{Pb_1}=2 mv.

As expected, an increase in the **learning constant c** leads to a quicker attainment of the steady regimen and to an augmentation of its dispersion. In addition, a decrease in the average amplitude of the synchronized response is observed, which however continues being periodic (Figure 5.24).

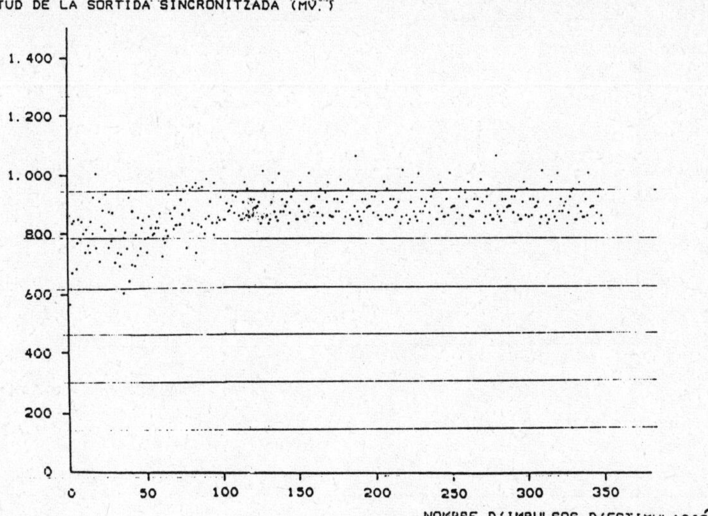

*** Figure 5.24 - Effect of the learning constant c: Evolution of the response synchronized with a stimulus of period 400 ms, during the process of learning this stimulus, for the same network used in the reference experiment, but with c=0.3.

5.3.2.3 - Initial state

From the accomplishment of the successive stages of the learning process in the same network with random initial state used in the reference experiment, but omitting the running in process, we can deduce the effect of the latter process upon subsequent learning. The comparison of graphs (a) and (b) in Figure 5.25 with those in Figures 5.6(a) and 5.8 seems to indicate the following trends: The amplitude of the response synchronized with the stimulus increases as a consequence of the running in process, while its temporal dispersion decreases, both before and after learning has taken place; the time until the attainment of the steady regimen remains unchanged. Other aspects such as the spatial distribution of spontaneous frequencies in the network, their position in the spectrum and the location of the regions where the learning of each frequency settles down predominantly, do not seem to be significantly affected by the running in process.

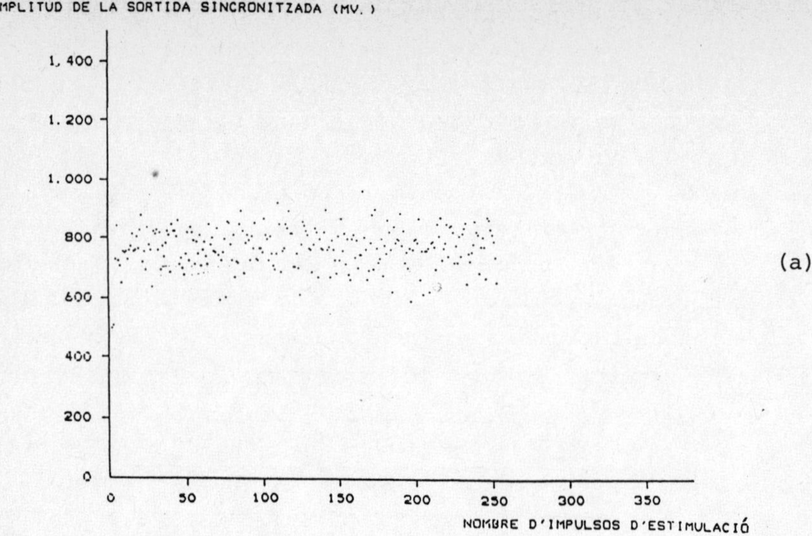

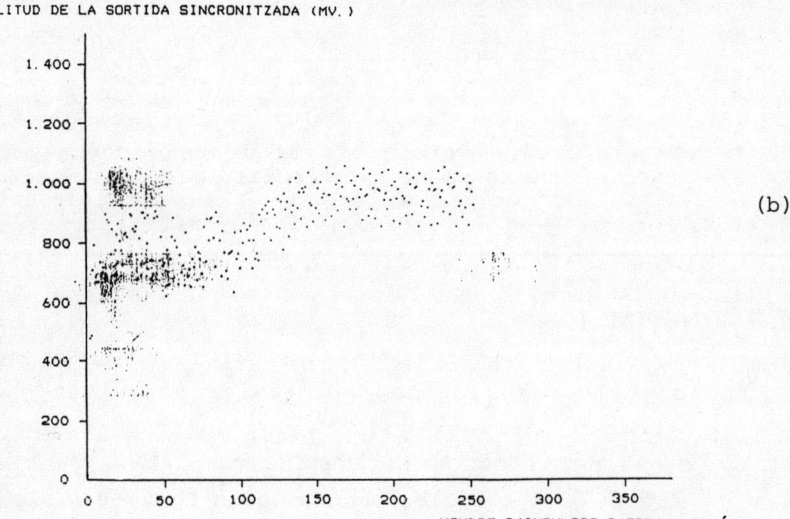

*** Figure 5.25 - Effect of the running in process: Evolution of the amplitude of the response synchronized with a stimulus of period 400 ms, before --(a)-- and after --(b)-- learning this stimulus, for the same network used in the reference experiment, but without its having undergone the running in process.

5.3.2.4 - Stimulation conditions

We will consider three factors relative to the stimulation conditions: the frequency of the stimulus, its amplitude and the sequence of presentation of the stimuli in the case of learning two frequencies.

From the simulations carried out with several networks, submitted to a wide gamut of periodic stimuli, it follows that the band of frequencies susceptible to being learned by neurons belonging to a network is slightly wider than the corresponding band for isolated neurons (see Section 3.3.4), fluctuating within the interval [1.15 Hz, 4 Hz], as a function of the values of the parameters n and p. This fact is explained by the shortening or lengthening of the firing period, with respect to the intrinsic one, provoked by the excitation or inhibition coming from other neurons. In general, the higher the **frequency of the stimulus**, the greater the number of neurons that learn it and the quicker learning takes place. This is reflected in the evolution of the response synchronized with the stimulus, which attains the steady regimen earlier, with a greater average amplitude and less dispersion through time (compare, for instance, Figures 5.8 and 5.14). These indications would not be conclusive if they were not confirmed by the bar diagram that characterizes the spontaneous behavior, where the exact number of neurons that fire at each frequency is reflected (e.g., Figures 5.9(c) and 5.15(d)).

The decisive role played by the bar diagram is made more evident when exploring the consequences of varying the **amplitude of the stimulus.** When the stimulus provokes a change greater than 20 mv in the membrane potential, practically all neurons in the network synchronize almost immediately either with the stimulus or with its first subharmonic (Figure 5.26), resulting in an evolution of the response synchronized with the stimulus with very little information content about the learning process. Then, it is necessary to make use of the index $\mathfrak{z}$ to determine the instant at which the steady regimen is attained and to use the bar diagram to determine the number of neurons that have actually learned the imposed frequency, distinguishing them from those that only synchronize with the stimulus. Through these indicators, one can observe that the learning process is basically the same for all stimuli with effect greater than 7 mv upon the unineuronal membrane potential (the bar diagram is practically equal to that in Figure 5.9(c)), being, however, slower and smaller in terms of the number of neurons, for stimuli with effect less than that amount.

Finally, the **sequence of presentation of the stimuli** affects the microstructure and the dispersion through time of the steady regimen attained by the response synchronized with the stimulus, for each learned frequency, modifying however neither its average amplitude, nor

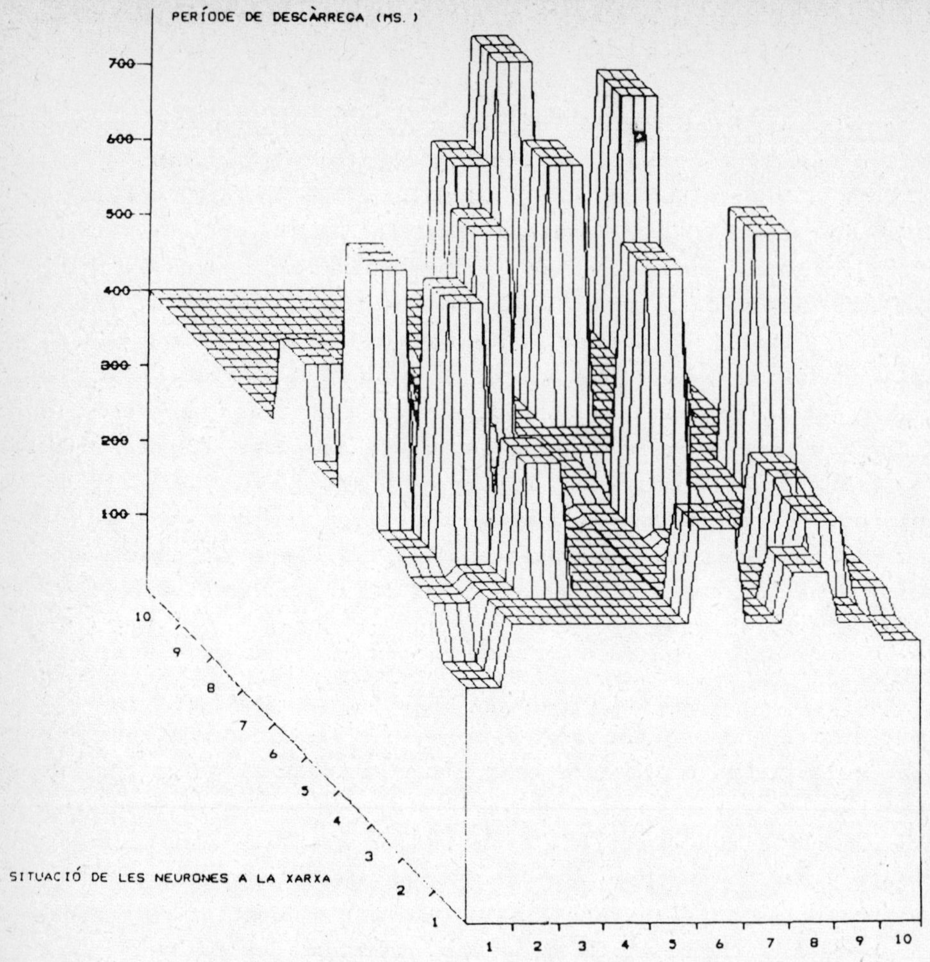

*** Figure 5.26 - Effect of the amplitude of the stimulus: Spatial distribution of frequencies in the network, while being submitted to a stimulus of period 400 ms and amplitude 25 mv.

the time spent until its attainment, nor the location in the network of the regions where the learning of each frequency predominates either. Figure 5.27 permits appreciating the described effects, by comparing its graphics with those in Figures 5.8, 5.14 and 5.15. Note that the process of learning the stimulus of period 400 ms always gives place to a periodic steady regimen, regardless of its being presented before or after the other stimulus (Figures 5.8 and 5.27(b)); while the periodicity observed when presenting the stimulus of period 550 ms in the second place (Figure 5.14) is induced by the previous steady regimen, since it does not appear when this is the first stimulus presented (Figure 5.27(a)).

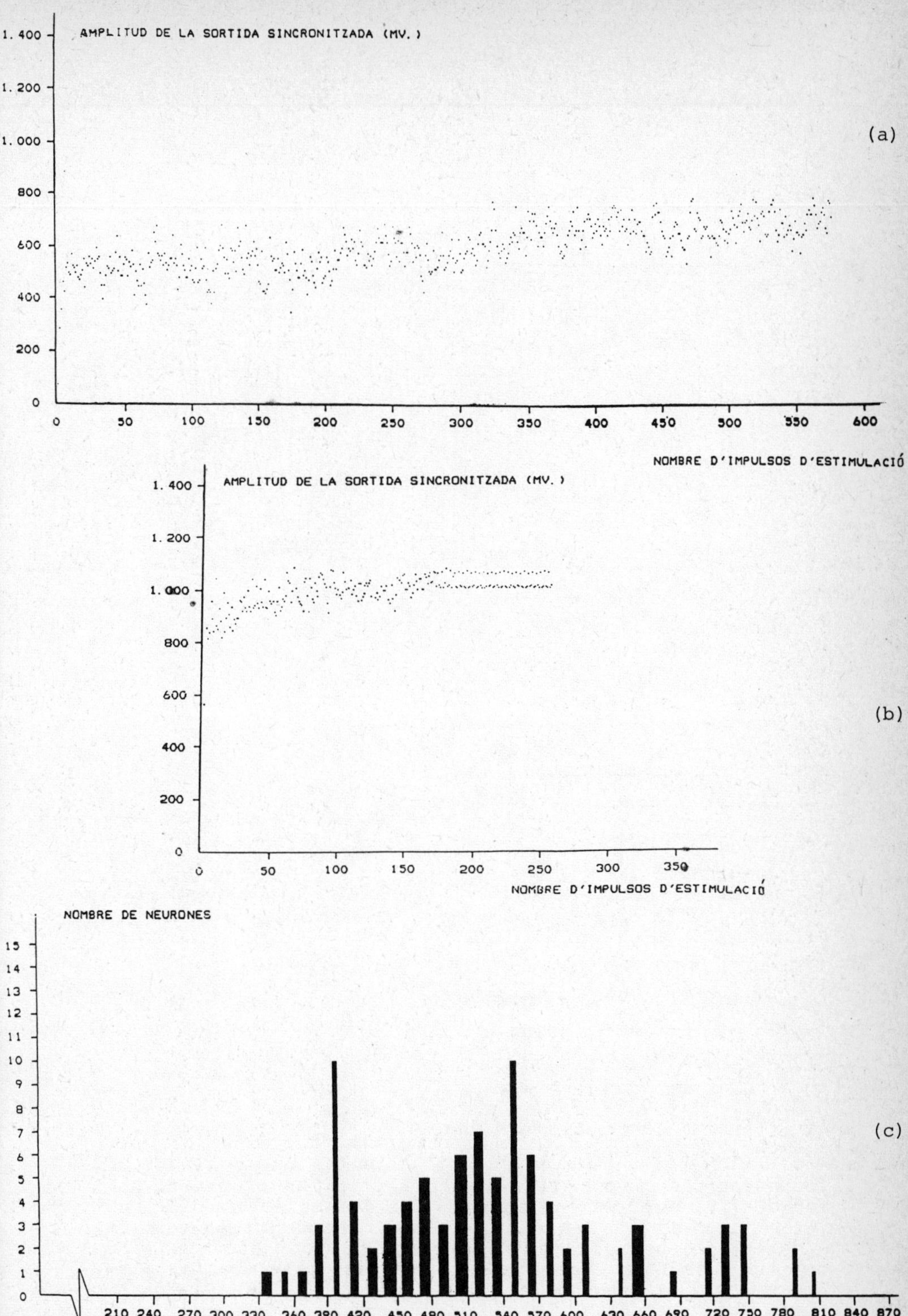
AMPLITUD DE LA SORTIDA SINCRONITZADA (MV.)
1.400
1.200
1.000
800
600
400
200
0
0 50 100 150 200 250 300 350 400 450 500 550 600
NOMBRE D'IMPULSOS D'ESTIMULACIÓ
(a)
AMPLITUD DE LA SORTIDA SINCRONITZADA (MV.)
1.400
1.200
1.000
800
600
400
200
0
0 50 100 150 200 250 300 350
NOMBRE D'IMPULSOS D'ESTIMULACIÓ
(b)
NOMBRE DE NEURONES
15 14 13 12 11 10 9 8 7 6 5 4 3 2 1 0
210 240 270 300 330 360 390 420 450 480 510 540 570 600 630 660 690 720 750 780 810 840 870
PERÍODE DE DESCÀRREGA (MS.)
(c)

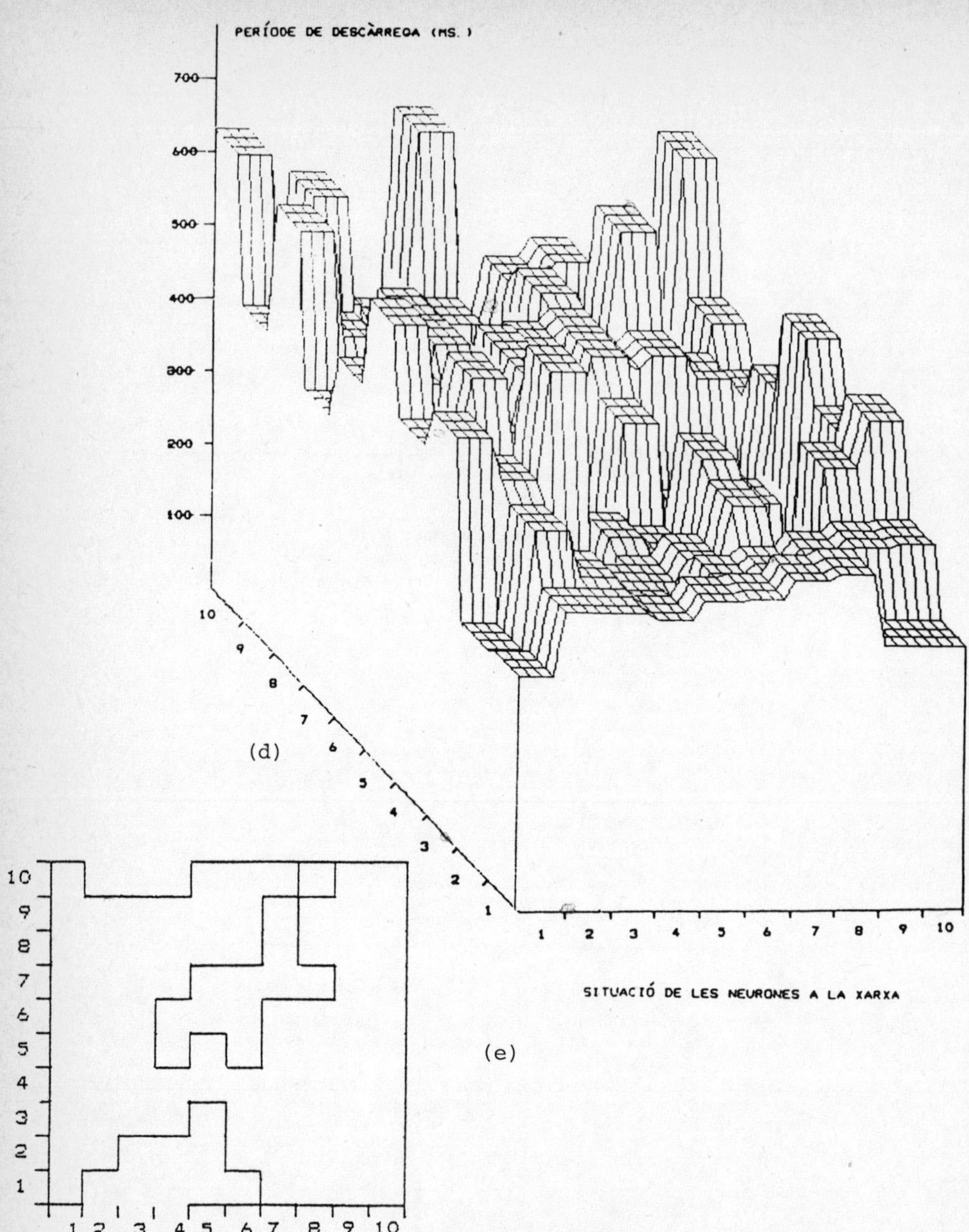

*** Figure 5.27 - Effect of reversing the order of presentation of the stimuli: Evolution of the amplitude of the response synchronized with a stimulus of period 550 ms --(a)-- and with a stimulus of 400 ms --(b)-- during the process of learning the two frequencies in this order; (c) bar diagram of average periods and standard deviations that characterizes the resulting spatial distribution of frequencies in the network --(d)--; and (e) demarcation of the regions where the learning of each frequency has settled down.

		SPONTANEOUS BEHAVIOR		EVOLUTION OF THE RESPONSE SYNCHRONIZED WITH THE STIMULUS*			OUTCOME OF THE LEARNING PROCESS			
		Activity level	Dispersion in the frequency-distribution	Time until the attainment of the steady state	Average amplitude	Temporal dispersion	Number of frequencies	Position in the spectrum	Clustering in regions	Location in the network
CONNECTIVITY	n ↑			↑		↓				
	p ↑	↑	↓		↑	↓	1 → 2 → 1	↔	↔	↔
	$\frac{a_e}{a_i}$ ↑		↓		↑		2 → 2 → 1		↔	
	Micro-connectivity									↔
	Excitation/inhibition overlap		↑						↔	↔
INTRANEURONAL FUNCTIONING	σ_{Pb_1} ↑			↑	↓	↑	2 → 0			
	c ↑			↓	↓	↑				
INITIAL STATE	Developmental process				↑	↑				
STIMULATION CONDITIONS	Stimulus frequency ↑			↓	↑	↓				
	Stimulus amplitude ↑			↓	↑	↓				
	Sequence of presentation of the stimuli					↔				

*** Table 5.4 - Summary of the effect of varying the several factors considered upon the behavior of the network. ↑= increase in the corresponding magnitude; ↓= decrease; ↔= change; 1 ⁀2⁀ 1 = evolution of the number of frequencies.

As promised at the beginning of this section, Table 5.4 provides a summary of the effect of the several factors considered upon the behavior of the network.

5.4 - CONCLUSIONS

In the present chapter, we have set up and studied through simulation a neural network model with 100 neurons, each represented by the model proposed in Chapter 3, interconnected according to a feedback lateral inhibition scheme. The fact that the neurons are pacemakers and plastic provides the network with a certain capacity of assimilating and reproducing external rhythms.

To explore the conditions in which the network reproduces the experimental phenomena described in Section 2.1.2, we have defined a set of measures and statistical procedures that permits detecting those phenomena and analyzing their characteristics. These measures and procedures are the following: index ξ of variation of the learning parameters m_{Pb_1}, curve of evolution of the response synchronized with a stimulus, spatial and/or temporal aggregation of spikes, spatial distribution of frequencies in the network, demarcation of the regions of uniform frequency, and bar diagram of average periods and standard deviations.

In an experiment, later used as reference, the network exhibited the following phenomena: progressive entrainment with a periodic stimulus, prediction of the instant of presentation of the next stimulation impulse, rhythm assimilation, generalization, discrimination and duplication of the stimulus in the firing pattern of some neurons. It is important to highlight the appearance of an asymmetry in the band of generalization of the learned frequencies, which extends more in the direction of the higher frequencies than in that of the lower ones; this asymmetry has neither been asserted nor denied in experimental studies, remaining thus as a prediction to be confirmed or refuted.

Taking the mentioned experiment as reference, we have studied the effect upon the behavior of the network of the variation of several factors, related to the connectivity, the intraneuronal functioning, the initial state and the stimulation conditions.

The factors that most crucially influence the learning capacity of the network are: the proportion of excitatory connections over the total (p), the ratio between the ranges of the excitatory and the inhibitory connectivity (a_e/a_i) and the degree of intraneuronal randomness (σ_{Pb_1}). The extreme values of p and the high values of (a_e/a_i) and σ_{Pb_1} reduce the number of frequencies susceptible to being learned to only one or, in the latter case, to none.

The former two of the above parameters, together with other factors linked to the connectivity, influence the dispersion of the initial distribution of frequencies in the network and determine whether the neurons that learn the same frequency cluster together and what the location in the network of the clusters is.

The remaining factors studied affect only the average amplitude, its dispersion through time and the time elapsed until the attainment of the steady regimen of the response synchronized with the stimulus.

In sum, the unineuronal learning rule proposed in Chapter 3, incorporated into a network model with feedback lateral inhibition, seems sufficient to reproduce many macroscopic phenomena of assimilation and endogenous generation of previously learned rhythms, observed in some experimental situations. The exploration carried out of the effect of varying several factors upon the behavior of the network qualifies the following lines of future research as being most promising: widening the gamut of frequencies susceptible to being learned by scaling the time-constants of the neuron model; systematizing the constraints imposed by the initial state upon the final spatial distribution of frequencies in the network; analytically studying the dynamic evolution of the network, using an extended version of the simplified model set up in Chapter 4 that includes the learning rule; and exploring the consequences of combining the temporal type of learning proposed in this thesis with the more classical ones of structural synaptic character.

CHAPTER 6
CONCLUSIONS AND FUTURE PROSPECTS

"I know why there are so many people who love chopping wood. In this activity one immediately sees the results".

A. Einstein

"En general puede decirse que no hay cuestiones agotadas, sino hombres agotados en las cuestiones".

S.Ramón y Cajal, 1897

The most outstanding results obtained, and the specific suggestions for future work in this line derived from them, have been detailed in the "Conclusions" section of each chapter. The present chapter attempts to put together these results, offering an overall view of the contribution of this monograph and its implications for future research.

6.1 - CONTRIBUTION OF THE RESEARCH ACCOMPLISHED

Following the outline set out in the introductory chapter, the main contributions can be summarized as follows:

(1) Design of a new neuron model (Chapter 3): A point of connection between two areas that until now had remained isolated --the biophysical modelling of intracellular activity and the study of adaptive neural networks-- has been established by proposing an electrophysiological model of pacemaker neurons provided with a biologically plausible mechanism to learn rhythms. The model overcomes the three limitations described in the Introduction that have traditionally constrained research in the second area. Specifically, the proposed learning rule, unsupervised and open-loop, provides an alternative to the classic structural Hebbian-like rules, that have not obtained the experimental support required to give neurobiological relevance to the profusion of theoretical studies to which they have given rise.

(2) Validation of the neuron model (Chapter 3): The model has shown ability to reproduce, through simulation, the set of experimental data previously defined as its domain of validity: the probability distribution of the spontaneous interspike interval, the phase response curve, entrainment intervals and plasticity in the firing pattern. This replicative validity has been complemented by a certain predictive capacity --in the case, for example, of stimuli of progressively increasing or decreasing frequency-- that suggests further experiments to corroborate or refute the suppositions on which the model is based, and that, in any case, will contribute to extending the knowledge about this type of neuron.

(3) Analytic study of the behavior of the model (Chapter 4): Characterizing the model by its PRC, the phase transition equation has been derived and, from it, the regions of the 3-D space of coordinates (initial phase, stimulation frequency and value of λ) where the different transitions take place and the infinite entrainment ratios are attained, have been determined. The set of stimulation frequencies that do not yield entrainment has turned out to be of measure zero, since the ratio between the stimulation and the firing frequencies has proven to be a generalized Cantor function of the ratio between the stimulation and the spontaneous frequencies. Taking advantage of the fact that the input/output pattern that emerges is unique for each entrainment ratio and that the set of these ratios is enumerable, an effective procedure for computing all the entrainment patterns has been developed. Thus, a full answer has been given to a question that had previously been left open (Segundo and Kohn, 1981).

The complete characterization of the entrainment regions and patterns generated by the deterministic model has permitted exploring the effect of randomness and learning. The first of these factors provokes systematic and consistent variations throughout the spectrum of patterns, which suggests a method of validating the adequacy of the model in different experimental situations that will be dealt with in the next section. In relation to learning, the analysis carried out indicates that it favors the appearance of simple entrainment ratios, since the regions where they occur become attractors.

(4) Specification of the network model (Chapter 5): A neural network containing a hundred neurons, randomly interconnected according to a probability distribution that materializes the principle of feedback lateral inhibition, has been defined. The fact that each neuron is represented by the model developed in this monograph provides the network with tentative mechanisms of generation of endogenous rhythms and of modification of its firing pattern, indispensable for recreating the experimental data that constitute the departure point of the modelling. The option taken, of incorporating autorhythmicity and plasticity at the level of the individual neuron, has been motivated by the theoretical interest of exploring new alternatives, as discussed in the introductory part of Chapter 5, and justified by the evidence given both at the neuron and network levels.

(5) Simulation of the network model (Chapter 5): To characterize the simulation results to facilitate comparison with experimental phenomena, a set of statistical treatments and specific measures has been defined that provides information about the degree of synchronization, convergence of learning, spatial distribution of frequencies in the network, and spatial and temporal dispersion of frequencies. The network has reproduced the following phenomena: progressive entrainment with a rhythmic stimulus, prediction of the instants of presentation of the stimulation impulses, assimilation of the rhythm, generalization, discrimination and duplication of the stimulus in the firing pattern of some neurons. From the exploration of the effect of varying several factors --related to the connectivity, the intraneuronal functioning, the initial state and the stimulation conditions-- upon the learning process, it follows that the most influential factors are: the proportion of excitatory connections over the total, the ratio between the ranges of the excitatory and the inhibitory connectivity, and the degree of intraneuronal randomness.

The research objective set forth in the Introduction seems fully achieved, since the proposed unineuronal learning rule has proved to reproduce, and so give an explanation for, the phenomena of assimilation and endogenous generation of previously-learned rhythms, observed when experimenting with isolated neurons of invertebrates and complete nervous systems of vertebrates. Thus, the work accomplished contributes to the research on the neuronal substratum of memory and learning of rhythms, with a new viewpoint that emphasizes the temporal dimension of the representation of information.

6.2 - AVENUES OF FUTURE RESEARCH

> *"Ya ves como el 'cogito', la Operación Humana por excelencia, se sitúa hoy en una región bastante vaga, entre electromagnética y química, y probablemente no se diferencia tanto como pensábamos de cosas tales como una aurora boreal (...) Sin contar -agregó Oliveira suspirando- que a lo mejor es al revés y resulta que la aurora boreal es un fenómeno espiritual y entonces sí que estamos como queremos...".*
>
> *J. Cortázar, 1963*

The neuron model has been structurally validated a priori and behaviorally validated a posteriori with regard to a set of experimental data initially defined as its particular reference framework. Still, the probabilistic analysis of the model, its simulation and the analytic study of its behavior imply a series of predictions that go beyond the mentioned framework and that require, for its confirmation or refutation, carrying out new experiments or applying specific statistical treatments (which will have to be designed ad hoc in some cases).

Some predictions refer to structure and others to behavior. Examples of the former type of prediction are the threshold values of the postsynaptic potential (P^*) and of the time elapsed since the last discharge (T^*), which determine respectively whether accelerative learning and whether decelerative learning occur. Concerning the latter type of prediction, we point out those related to the effect of randomness upon the PRC and upon the input/output patterns originated by rhythmic stimulation. This effect probably constitutes the line of future research most directly approachable from the results obtained in the present work.

The research lines described are basically experimental and require systematization of the data and elaboration of statistical tests to determine the adequacy of the model to the data obtained. In Chapters 3 and 4, experimental and theoretical designs have been suggested that would permit extending both structural and behavioral predictions, thus contributing to an ulterior validation of the model and opening, for the latter type of predictions, a new way of validating the aspects linked to randomness of any neuronal oscillator model.

Another avenue opened by the present work that, if the tendency observed in the simulations carried out is confirmed, could have great relevance within the experimental domain, is the one that refers to the effect of learning upon the entrainment patterns. This avenue, still in an embryonic stage, would require an analytic study like the one done in Chapter 4, but this time based on an extension of the simplified model there proposed that would incorporate the learning rule. The difficulty is that learning entails the modification of the parameters of the model, which then becomes time-variant. The importance of this avenue is made more evident by noting that its logical continuation would be the analytic study of the dynamic evolution of the network.

Another theoretical avenue was pointed out in Chapter 4, when mentioning the experimental interest of analyzing the implications of introducing certain qualitative modifications in the piecewise linear PRC studied. The procedure would be to study the sensitivity of the model to these changes and perhaps, in the final analysis, to arrive at a more generic simplified model, adjustable to different experimental preparations.

While the most promising avenues of future research on the neuron model are experimental or mathematical, in the case of the network model there are still many open questions that need to be approached through simulation. At the end of Chapter 5, the most interesting of these open questions, in the light of the exploration carried out on the effect of different factors upon the behavior of the network, have been listed. Briefly, they refer to the repercussion of scaling the time-constants of the neuron model, of varying the initial state of the network, and of combining the proposed temporal-type learning with the more classical one of synaptic structural character. Among them, the last is undoubtedly the most important.

APPENDIX A
NEURONAL PHYSIOLOGY

In this appendix, we outline the basic concepts of neurophysiology assumed known when describing in Chapter 2 the phenomena that we attempt to model.

A more detailed description of the concepts, as well as exceptions to almost all the generic principles that we will enunciate, are found in Ganong (1966), Guyton (1973), Shepherd (1974), Eccles (1977), Lund (1978) and Stevens (1979).

Since Ramón y Cajal (1899) ended the controversy over the protoplasmatic continuity of the nervous system, establishing the existence of separate nervous cells, the conceptualization of this system as a network of neurons has led neurophysiological research.

STRUCTURE OF THE NEURON

Neurons adopt a great variety of forms, according to where they are found within a nervous system and their species. However, they also share certain structural characteristics that make it possible to distinguish in them three regions (Figure A.1):

(a) The **dendrites** (from the Greek, "dendron", tree) are ramifications that provide the main surface through which the neuron receives input signals.

(b) The **soma** contains the nucleus of the neuron and the biochemical machinery for the synthesis of proteins.

(c) The **axon** is the fibre that conducts the impulses generated by the neuron toward other parts of the nervous system. Its final portion divides into branches, each ending in a small bulb that affects the dendrites of other neurons.

The direction of propagation of the signals is always from the dendrites toward the axon.

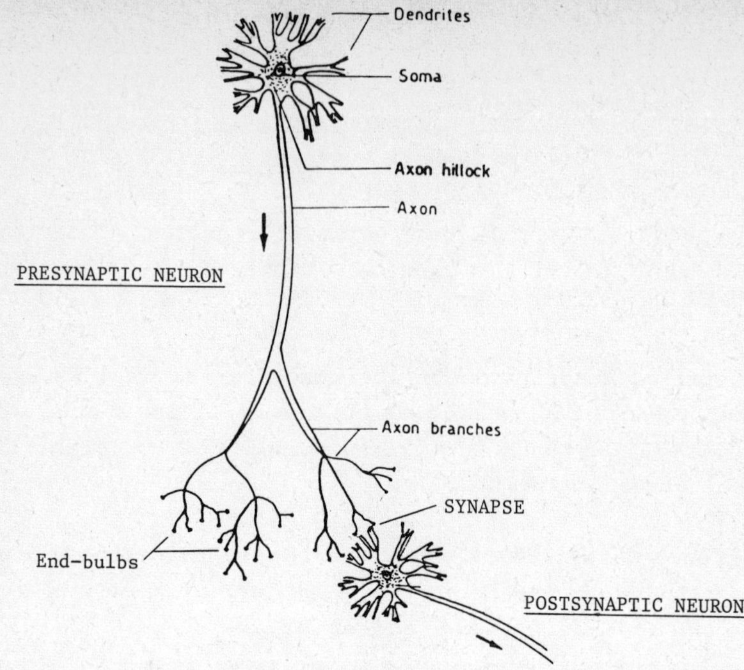

*** Figure A.1 - Structure of the neuron and of the synapses.

MEMBRANE POTENTIAL AND ACTION POTENTIAL

In the resting state, there is a difference of potential between the inside and the outside of the cellular membrane, caused by differences in the concentration of sodium and potassium ions. In non-pacemaker neurons, this resting potential is about -70 mv and any change in it, caused by external stimulation or stimulation coming from another neuron, is passively propagated in a progressively-damped way through the dendrites and the soma.

In pacemaker neurons, the membrane potential not only varies because of the above reasons, but also because of endogenous processes related to the opening rhythm of the channels and the functioning of the pumps, which together regulate the flow of ions through the membrane. The mentioned channels and pumps, also present in non-pacemaker neurons, are activated by specific proteins and account for the selective permeability of the membrane to the different ions: the channels in the direction determined by the difference of potential, and the pumps in the direction of overcoming its resistance.

If the difference of potential at the origin of the axon (the axon hillock) surpasses a certain threshold, then an impulse is generated that actively propagates along the full length of the axon to arrive at the end-bulbs. The autoregenerative difference of potential underlying the mentioned active propagation is called the action potential and the generated impulse, a discharge or spike.

The action potential is originated by changes in the permeability of the membrane to the different ions, which requires certain proteins to open selectively the channels corresponding to the various ions. After a discharge occurs, there is a refractory period in which the axon cannot actively propagate another impulse, since the proteins associated to the channels have adopted an inactive configuration different from the one characteristic of the resting state. The refractory period is at first absolute, later becoming relative, in the sense that a large enough difference of potential can cause a new discharge.

SYNAPTIC TRANSMISSION

The synapse is the place where an end-bulb of the axon of a neuron, called presynaptic, acts on the dendrites of another neuron, called postsynaptic (Figure A.1). In the synapse, the interneuronal transmission of information takes place not electrically, but chemically.

The arrival of an impulse at the end-bulb causes the liberation of the neurotransmitter contained in its vesicles, which spreads through the synaptic cleft and finally provokes a change in the membrane potential of the postsynaptic neuron. Depending on whether a depolarization or a hyperpolarization occurs, the synapse and the postsynaptic potential generated are called excitatory or inhibitory.

The fact that the synapse is excitatory or inhibitory does not depend as much on the liberated neurotransmitter, as on the receptor that is in the postsynaptic neuron. A single neuron can thus give place to both excitatory and inhibitory synapses.

The postsynaptic potentials generated in the different synapses of a neuron propagate passively through their dendrites and soma, according to what was expressed above, making the spatial and temporal summation of the influences that come from the different presynaptic neurons possible.

In Figure A.2, the temporal evolution of the different potentials described is represented.

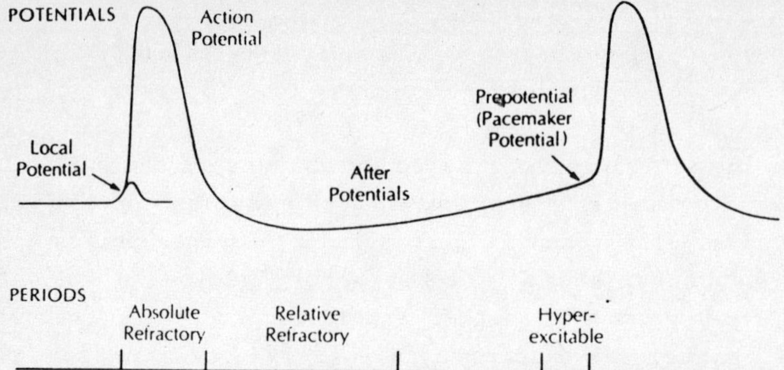

*** Figure A.2 - Evolution of the spontaneous potential, the postsynaptic potential and the action potential (from Shepherd, 1974).

APPENDIX B

PROPERTIES OF MAPPINGS FROM THE UNIT CIRCLE ONTO ITSELF

A detailed treatment of the concepts and results described here can be found in Denjoy (1932), Coddington and Levinson (1955), Brunovský (1974) and Herman (1977).

Let $f: S^1 \dashrightarrow S^1$ be a continuous onto mapping which preserves the orientation of the circle and such that:

$$f(x_1) = f(x_2) \Rightarrow f(x_1) = f(x), \; \forall x \in [x_1, x_2]$$

then the following results hold:

PROPOSITION B.1. The mapping $\hat{f}: \mathbb{R} \dashrightarrow \mathbb{R}$ defined by:

$$\hat{f}(x+n) = \begin{cases} f(x)+n, & \text{if } f(x) \geq f(0) \\ f(x)+n+1, & \text{otherwise} \end{cases} \qquad \forall x \in S^1, \; \forall n \in \mathbb{Z}$$

is continuous, onto and monotonically increasing.

PROPOSITION B.2. The limit:

$$\rho = \lim_{|n| \to \infty} \frac{\hat{f}^n(x)}{n}$$

exists and is independent of x.

The number ρ, called **the rotation number** of f, measures the average advancement in $\mathbb{R}$ produced by each iteration of f.

PROPOSITION B.3. $\rho \in \mathbb{Q} \Longleftrightarrow \exists n \in \mathbb{N} \; \wedge \; \exists \emptyset^* \in S^1$ s.t. $f^n(\emptyset^*) = \emptyset^*$.

We will write (x,y), with $x,y \in \mathbb{N}$, to denote the greatest common divisor of x and y.

PROPOSITION B.4. If $\rho = (r/s)$ with $r,s \in \mathbb{N}$, $s \neq 0$ and $(r,s)=1$, then:

$$\left[f^n(\emptyset^*) = \emptyset^* \Longrightarrow (n,s) = s \;\wedge\; \hat{f}^s(\emptyset^*) = \emptyset^* + r\right]$$

If the mapping f depends on a parameter $p \in P$, P being a one-dimensional variety,

$$f : P \times S^1 \longrightarrow S^1$$
$$(p, x) \longmapsto f_p(x)$$

then we actually have a family of mappings $\{f_p\}_{p \in \mathbb{R}^+}$ of the type considered, each having a rotation number ρ_p associated with it.

We can define the mapping:

$$\rho_f : \mathbb{R}^+ \longrightarrow \mathbb{R}$$
$$p \longmapsto \rho_p$$

PROPOSITION B.5. If $\hat{f}: \quad P \times \mathbb{R} \longrightarrow \mathbb{R}$

$$(p, x) \longmapsto \hat{f}_p(x)$$

is continuous and monotonically increasing with p, then so is ρ_f.

APPENDIX C

GLOSSARY OF CATALAN LABELS IN COMPUTER-MADE FIGURES

Amplitud de la resposta: Output amplitude.

Amplitud de la sortida sincronitzada: Amplitude of the synchronized output.

Avançament de fase: Phase-advancement.

Estímul: Stimulus.

Nombre de neurones: Number of neurons.

Nombre d'impulsos d'estimulació: Number of stimulation impulses.

Període de descàrrega: Firing period.

Període de l'estimulació: Period of the stimulation.

Potencial de membrana: Membrane potential.

Retardament de fase: Phase-delay.

Situació de les neurones a la xarxa: Location of neurons in the network.

REFERENCES

Albus, J.S. (1971): "A theory of cerebellar function". Math. Biosci. 10, 25-61.

------------ (1979): "Mechanisms of planning and problem solving in the brain". Math. Biosci. 45, 247-293.

Alkon, D.L., I. Lederhendler and J.J. Soukimas (1982): "Primary changes of membrane currents during retention of associative learning". Science 215, num. 4533.

Amari, S. (1977a): "A mathematical approach to neural systems". In "Systems Neuroscience", edited by J. Metzler, Academic Press, New York.

---------- (1977b): "Neural theory of association and concept-formation". Biol. Cybern. 26, 175-185.

---------- (1977c): "Dynamics of pattern formation in lateral-inhibition type neural fields". Biol. Cybern. 27, 77-87.

---------- (1980): "Topographic organization of nerve fields". Bull. of Math. Biol. 42, 339-364.

---------- and M.A. Arbib (1977): "Competition and cooperation in neural nets". In "Systems Neuroscience", edited by J. Metzler, Academic Press, New York.

Andersen, P. and S.A. Andersson (1968): "Physiological basis of the alpha rhythm". Appleton-Century-Crofts, New York.

------------- and J.C. Eccles (1962): "Inhibitory phasing of neuronal discharge". Nature 196, 645-647.

------------- , ----------- and T.A. Sears (1964): "The ventro-basal complex of the thalamus: types of cells, their responses and their functional organization". J. Physiol. 174, 370-399.

Anderson, J.A., J.W. Silverman, S.A. Ritz and R.S. Jones (1977): "Distinctive features, categorical perception and probability learning: Some applications of a neural model". Psychol. Rev. 85, 413-451.

Andronov, A.A., A.A. Vitt and S.E. Khaikin (1966): "Theory of Oscillators". Pergamon, Oxford.

Arbib, M.A. (1972): "The Metaphorical Brain". J. Wiley & Sons.

----------, C.C. Boylls and P. Dev (1974): "Neural models of spatial perception and the control of movement". In "Kybernetik und Bionik/ Cybernetics and Bionics", edited by W.D. Keidel, W. Handler and M. Spreng, Munich/Vienna, Oldenbourg.

----------, W.L. Kilmer and D.N. Spinelli (1976): "Neural Models and Memory". In "Neural Mechanisms of Learning and Memory", edited by M.R. Rozenzweig and E.L. Bennet, MIT Press.

Ayers, J.L. and A.I. Selverston (1977a): "Synaptic control of an endogenous pacemaker network". J. Physiol. (Paris) 73, 453-461.

----------- and ---------------(1977b): "Monosynaptic control of inter- and intra-oscillator coordination of an endogenous pacemaker network". Soc. Neurosci. Abstr. 7, 267.

----------- and --------------- (1979): "Monosynaptic entrainment of an endogenous pacemaker network: a cellular mechanism for von Holt's magnet effect". J. Comp. Physiol. A 129, 5-17.

Barker, J.L. and T.S. Smith (1978): "Electrophysiological studies of molluscan neurons generating bursting pacemaker potential activity". In "Abnormal Neuronal Discharges", edited by N. Chalazonitis and M. Boisson, Raven Press.

Barto, A.G. and R.S. Sutton (1981a): "Landmark learning: An illustration of associative search". Biol. Cybern. 42, 1-8.

---------- and ---------- (1981b): "Goal-seeking components for adaptive intelligence: An initial assessment". Tech. Report Avionics Lab., Air Force Wright Aeronautical Laboratories, Wright-Patterson Air Force Base, Ohio.

---------- and ---------- (1982): "Simulation of anticipatory responses in classical conditioning by a neuron-like adaptive element". Behav. Brain Res. 4, 221-235.

---------- and ---------- (1983): "Neural problem solving". University of Massachusetts, COINS Tech. Report 83-03.

---------- , ---------- and Ch.W. Anderson (1983): "Neuron-like adaptive elements that can solve difficult learning control problems". IEEE Trans. on Syst., Man and Cybern., SMC-13, num. 5, 834-846.

---------- , ---------- and P.S. Brouwer (1981): "Associative Search Network: A reinforcement learning associative memory". Biol. Cybern. 40, 201-211.

Beltz, B. and A. Gelperin (1980): "Mechanisms of peripheral modulation of salivary burster in Limax maximus: a presumptive sensorimotor neuron". J. Neurophysiol. 44, 675-686.

Bernussou, J. (1977): "Point mapping stability". Pergamon Press, Oxford.

Best, E.N. (1979): "Null space in the Hodgkin-Huxley equations: a critical test point". Biophys. J. 27, 87-104.

Bonhoeffer, K.F. (1948): "Activation of passive iron as a model for excitation in nerve". J. Gen. Physiol. 32, 69-91.

Borges, J.L. (1954): "Del rigor en la ciencia". In "Historia universal de la infamia". Emecé, Buenos Aires.

Brindley, G.S. (1969): "Nerve net models of plausible size that perform many simple learning tasks". Proc. of the Royal Soc. (London) B 174, 173-191.

Bristow, D.G. and J.W. Clark (1982): "A mathematical model of primary pacemaking cell in a SA node of the heart". The American Physiological Society.

Brooks, V.B. and V.J. Wilson (1959): "Recurrent inhibition in the cat's spinal cord". J. Physiol. 146, 380-391.

Brunovský, P. (1974): "Generic properties of the rotation number of one-parameter diffeomorphisms of the circle". Czechoslovak Math. J. 24 (99), 74.

Bullock, T.H. (1974): "Comparisons between vertebrates and invertebrates in nervous organization". In "The Neurosciences: Third Study Program", edited by F.O. Schmitt and F.G. Worden, MIT Press, 343-346.

------------ (1976): "In search of principles in neural integration". In "Simpler Networks and Behavior", edited by J.C. Fentress, Sinauer Associates, Massachusetts.

Bunge, M. (1981): "From Mindless Neuroscience and Brainless Psychology to Neuropsychology". Whiting lecture: Winter Conference on Brain Research, Colorado.

Buño, W. and J. Fuentes (1984): "Coupled oscillators in an isolated pacemaker-neuron?". Brain Res. 303, 101-107.

-------- and ---------- (1985): "Resetting in crayfish stretch receptor". J. Physiol., to appear.

Buser, P. (1976): "Higher functions of the nervous system". Ann. Rev. Physiol. 38, 217-245.

Bustamante, J. (1980): "Receptores sensoriales periféricos". Memoria de la investigación realizada en el Servicio de Neurología experimental, Departamento de Investigación del Centro Especial "Ramón y Cajal" de la Seguridad Social de Madrid.

------------- , J. Fuentes and W. Buño (1981): "Análisis intracelular de los factores que influyen en la irregularidad de un marcapasos". FESBE 2, Madrid.

Calvin, W.H. (1974): "Three modes of repetitive firing and the role of threshold time course between spikes". Brain Res. 69, 341-346.

Coddington, E.A. and N. Levinson (1955): "Theory of ordinary differential equations". McGraw-Hill, New York.

Cohen, M.J. (1970): "A comparison of invertebrate and vertebrate central neurons". In "The Neurosciences: Second Study Program", edited by F.O. Schmitt, Rockefeller University Press, 798-812.

Cortázar, J. (1963): "Rayuela". Ed. Sudamericana, Buenos Aires.

------------ (1967): "La vuelta al día en 80 mundos". Siglo XXI, México.

Creutzfeldt, O.D., S. Watanabe and H.D. Lux (1966): "Relations between EEG phenomena and potentials of single cortical cells. I-Evoked responses after thalamic and epicortical stimulation". Electroencephalography and Clin. Neurophysiol. 20, 1-18.

Denjoy, A. (1932): "Sur les courbes définies par les équations différentielles à la surface du tore". J. Math. Pures Appl. (9) 11, 333-375.

Dev, P. (1975): "Computer simulation of a dynamic visual perception model". Int. Journ. of Man-Machine Studies 7, 511-528.

Didday, R.L. (1970): "The simulation and modelling of distributed information processing in the frog visual system". Ph.D. Thesis, Stanford University.

------------- (1976): "A model of visuomotor mechanisms in the frog optic tectum". Math. Biosci. 30, 169-180.

Duda, R.O. and P.E. Hart (1973): Pattern classification and scene analysis". Wiley, New York.

Eberly, L., M. Lasseter and H. Pinsker (1979): "Phase response curves and entrainment of the interneuron II oscillator in intact behaving Aplysia". Neurosci. Abstr. 5, 244.

Eccles, J.C. (1977): "The understanding of the brain". McGraw-Hill.

------------ , P. Fatt and K. Koketsu (1954): "Cholinergic and inhibitory synapses in a pathway from motor-axon collaterals to motoneurons". J. Physiol. 126, 524-562.

Enright, J.T. (1980): "Temporal precision in circadian systems: A reliable clock from unreliable components?". Science 209, 1542-1545.

Ernst, D., B. Bargel and F. Holdermann (1976): "Processing of remote sensing data by a region growing algorithm". Proc. 3rd Int. Joint Conf. on Pattern Recognition, California.

Feldman, J.L. and C.L. Cleland (1982): "Possible roles of pacemaker neurons in mammalian respiratory rhythmogenesis". In "Cellular Pacemakers", edited by D.O. Carpenter, J. Wiley & Sons, New York, 104-128.

Fidone, S.J. and J.B. Preston (1971): "Inhibitory resetting of resting discharge of fusimotor neurons". J. Neurophysiol. 34, 217-227.

Firth, D.R. (1966): "Interspike interval fluctuations in the crayfish stretch receptor organ". Biophys. J. 6, 201-215.

Fitzhugh, R. (1961): "Impulses and physiological states in theoretical models of nerve membrane". Biophys. J. 1, 445-466.

Fohlmeister, J.F., R.E. Poppele and R.L. Purple (1974): "Repetitive firing: Dynamic behavior of sensory neurons reconciled with a quantitative model". J. Neurophysiol. 37, 1213-1227.

------------------, ------------- and ----------- (1977): "Repetitive firing: A quantitative study of feedback in model encoders". J. Gen. Physiol. 69, 815-848.

Frazier, W., R. Waziri and E.R. Kandel (1965): "Alterations in the frequency of spontaneous activity in Aplysia neurons with contingent and non-contingent nerve stimulation". Fed. Proc. 24, 522.

Fuentes, J. (1979): "Contribución al procesamiento digital de señales biológicas". Tesis doctoral, E.T.S.I. de Telecomunicación, Universidad Politécnica de Madrid.

Gäwiler, B.H. and J.J. Dreifuss (1979): "Phasically firing neurons in long-term cultures of the rat hypothalamic supraoptic area: pacemaker and follower cells". Brain Res. 177, 95-103.

Ganong, W.F. (1966): "Manual de Fisiología Médica". El Manual Moderno S.A., México.

Gelbaum, B.R. and J.M.H. Olmsted (1964): "Counterexamples in Analysis". Holden-Day, San Francisco.

Gillette, R., M.V. Gillette and W.J. Davis (1980): "Action potential broadening and endogenously sustained bursting are substrates of command ability in a feeding neuron of Pleurobranchaea". J. Neurophysiol. 43, 669-685.

Glaser, E.R. and D.S. Ruchkin (1976): "Principles of Neurobiological Signal Analysis". Academic Press.

Glass, L., C. Graves, G.A. Petrillo and M.C. Mackey (1980): "Unstable dynamics of a periodically driven oscillator in the presence of noise". J. Theor. Biol. 86, 455-475.

--------- and M.C. Mackey (1979): "A simple model for phase locking of biological oscillators". J. Math. Biol. 7, 339-352.

Grossberg, S. (1972): "Neural expectation: Cerebellar and retinal analogs of cells fired by learnable or unlearned pattern classes". Kybernetic 10, 49-57.

------------- (1974): "Classical and instrumental learning by neural networks". In "Progress in theoretical biology", edited by R. Rosen and F. Snell, Academic Press, New York.

Guckenheimer, J. (1977): "On the bifurcation of maps of the interval". Inventions Math. 39, 165-178.

Guevara, M.R. and L. Glass (1982): "Phase locking, period doubling bifurcations and chaos in a mathematical model of a periodically driven oscillator: A theory of the entrainment of biological oscillators and the generation of cardiac disrhythmias". J. Math. Biol. 14, 1-3.

----------- , ---------- and A. Shrier (1981): "Phase locking, period-doubling bifurcations, and irregular dynamics in periodically stimulated cardiac cells". Science 214, 18.

Guttman, R., S. Lewis and J. Rinzel (1980): "Control of repetitive firing in squid axon membrane as a model for a neuroneoscillator". J. Physiol. 305, 377-395.

Guyton, A.G. (1973): "Fisiología Humana". Ed. Interamericana, México.

Harding, G.W. and A.L. Towe (1978): "What we need is a smart electrode". Brain Theory Newsletter 3, num. 3/4.

Harmon, L.D. and E.R. Lewis (1966): "Neural Modelling". Physiol. Rev. 46, 513-591.

Harth, E. and E. Tzanakou (1974): "ALOPEX: a stochastic method for determining visual receptive fields". Vision Res. 14, 1475-1482.

Hartline, D.K. (1976a): "Simulation of phase-dependent pattern changes to perturbations of regular firing in crayfish stretch receptor". Brain Res. 110, 245-257.

------------- (1976b): "SNAX: A language for interactive neuronal modelling and data processing". In "Computer Technology in Neuroscience", edited by P.B. Brown, Hemisphere.

------------- (1979): "Pattern generation in the lobster (Panulirus) stomatogastric ganglion. II - Pyloric network simulation". Biol. Cybern. 33, 223-236.

------------- and D.V. Gassie (1979): "Pattern generation in lobster (Panulirus) stomatogastric ganglion. I - Pyloric neuron kinetics and synaptic interactions". Biol. Cybern. 33, 209-222.

Hartline, H.K., F. Ratliff and W.H. Miller (1961): "Inhibitory interaction in the retina and its significance in vision". In "Nervous Inhibition", edited by E. Florey, Pergamon Press.

Hebb, D.O. (1949): "The organization of behavior". Wiley, New York.

Herman, M.R. (1977): "Measure de Lebesgue et nombre de rotation". In "Geometry and Topology", edited by J. Palis and M. de Carmo, Lecture Notes in Math. 597, 271-293, Springer-Verlag.

Hodgkin, A.L. and A.F. Huxley (1952): "A quatitative description of membrane current and its application to conduction and excitation in nerve". J. Physiol. 117, 500-544.

Hofstadter, D.R. (1981): "A conversation with Einstein's brain". In "The Mind's I", Bantam Books.

Holden, A.V. and S.M. Ramadan (1980): "Identification of endogenous and exogenous activity in a molluscan neurone by spike train analysis". Biol. Cybern. 37, 107-114.

----------- and ----------- (1981a): "The response of a molluscan neurone to a cyclic input: entrainment and phase-locking". Biol. Cybern. 41, 157-163.

----------- and ----------- (1981b): "Repetitive activity of a molluscan neurone driven by maintained currents: A supercritical bifurcation". Biol. Cybern. 42, 79-85.

Hoppensteadt, F.C. and J.P. Keener (1982): "Phase locking of biological clocks". J. Math. Biol. 15, 339-349.

Hubel, D.H. and T.N. Wiesel (1959): "Receptive fields of single neurons in the cat's striate cortex". J. Physiol. 148, 574-591.

---------- and ---------- (1963): "Receptive fields of cells in striate cortex of very young, visually inexperienced kittens". J. Neurophysiol. 26, 994-1002.

Ikeda, N. (1982): "Model of bidirectional interaction between myocardial pacemakers based on the Phase Response Curve". Biol. Cybern. 43, 157-167.

---------- , H. Tsuruta and T. Sato (1981): "Difference equation model of the entrainment of myocardial pacemaker cells based on the Phase Response Curve". Biol. Cybern. 42, 117-128.

Ito, M. (1982): "Mechanisms of motor learning". In "Competition and Cooperation in Neural Nets", edited by S. Amari and M.A. Arbib, Lecture Notes in Biomath. 45, Springer-Verlag.

Jalife, J. and Ch. Antzelevitch (1979): "Phase resetting and annihilation of pacemaker activity in cardiac tissue". Science 206, num. 9.

John, E.R. (1967): "Mechanisms of memory". Academic Press.

---------- (1972): "Switchboard versus statistical theories of learning and memory". Science 177, 850-864.

---------- and E.L. Schwartz (1978): "The neurophysiology of information processing and cognition". Ann. Rev. Psychol. 29, 1-29.

Junge, D. and G.P. Moore (1966): "Interspike-interval fluctuations in Aplysia pacemaker neurons". Biophys. J. 6, 411-434.

Kandel, E.R. (1967): "Cellular studies of learning". In "The Neurosciences: A Study Program", 666-689.

------------ (1976): "Cellular basis of behavior: An introduction to behavioral neurobiology", Freeman.

------------ (1978): "A cell-biological approach to learning". Grass Lecture Monograph 1, Soc. Neurosc., Bethesda.

------------ and W.A. Spencer (1968): "Cellular neurophysiological approaches in the study of learning". Physiol. Rev. 48, 65-134.

Karplus, W.J. (1983): "The spectrum of mathematical models". Perspectives in Computing 3, num. 2, 4-13.

Keener, J.P. (1980): "Chaotic behavior in piecewise continuous difference equations". Trans. Am. Math. Soc. 261, num. 2, 589-604.

------------ (1981): "On cardiac arhythmias: AV conduction block". J. Math. Biol. 12, 215-225.

------------ , F.C. Hoppensteadt and J. Rinzel (1981): "Integrate-and-fire models of nerve membrane response to oscillatory input". SIAM J. Appl. Math. 41, num. 3, 503-517.

Kilmer, W.L. and M. Olinski (1974): "Model of a plausible learning scheme for CA3 hippocampus". Kybernetic 16, 133-144.

Klopf, A.H. (1972): "Brain function and adaptive systems --a heterostatic theory". Air Force Cambridge Res. Labs, Res. Report AFCRL-72-0164 (A summary in: Proc. IEEE Int. Conf. Syst., Man and Cybern., Texas, 1974).

Knight, B.W. (1972): "Dynamics of encoding in a population of neurons". J. Gen. Physiol. 59, 734-766.

Kohn, A.F. (1980): "Influence of presynaptic irregularity on the inhibition of a pacemaker in crayfish and neuromine". Ph.D. Thesis, Univer. of California.

---------- , A. Freitas da Rocha and J.P. Segundo (1981): "Presynaptic irregularity and pacemaker interactions". Biol. Cybern. 41, 5-18.

Kohonen, T. (1977): "Associative memory: A system theoretic approach". Springer-Verlag, Berlin.

----------- (1982): "Self-organized formation of topologically correct feature maps". Biol. Cybern. 43, 59-69.

----------- and E. Oja (1976): "Fast adaptative formation of orthogonalizing filters and associative memory in recurrent networks of neuron-like elements". Biol. Cybern. 21, 85-95.

Krasne, F.B. (1978): "Extrinsic control of intrinsic neuronal plasticity: A hypothesis from work on simple systems". Brain Res. 140, 197-216.

Kristan, W.B. (1971): "Plasticity of firing patterns in neurons of Aplysia pleural ganglion". J. Neurophysiol. 34, 321-336.

-------------- (1980): "Generation of rhythmic motor patterns". In "Information Processing in the Nervous System", edited by H.M. Pinsker and W.D. Willis, Raven Press.

-------------- and R.L. Calabrese (1976): "Rhythmic swimming activity in neurons of the isolated nerve cord of the leech". J. Exp. Biol. 65, 643-668.

Lara, R., R. Tapia, F. Cervantes, A. Moreno and H. Trujillo (1980): "Mathematical models of synaptic plasticity: I - Post-tetanic potentiation; II - Habituation; III - Heterosynaptic facilitation". Neurol. Res. 1, num. 4.

-------- , --------- , C. Romero and F. Soria (1976): "Electronic simulation of synaptic facilitation and conditioning". 8e Congrès Int. Cybern., Namur.

Leiman, A.L. and C.N. Christian (1973): "Electrophysiological basis of learning and memory". In "The Physiological Basis of Memory", edited by J.A. Deutsch, Academic Press.

Lettwin, J.Y., H.R. Maturana, W. Pitts and W.S. McCulloch (1960): "Two remarks on the visual system of the frog". In "Sensory Communication", edited by W.A. Rosenblith, John Wiley & Sons., New York.

Levine, D.S. (1984): "Neural Population Modeling and Psychology: A Review". Math. Dept., Univ. Houston, in press.

Li, T.Y. and J.A. Yorke (1975): "Period three implies chaos". Am. Math. Monthly 82, 985-992.

Livanov, M.N. and K.L. Poliakov (1945): "The electrical reactions of the cerebral cortex of a rabbit during the formation of a conditioned defense reflex by means of rhythmic stimulation". Izvestiya Akademiya Nauk, USSR Ser. Biol. 3, 286.

Lund, R.D. (1978): "Development and plasticity of the brain". Oxford Univ. Press.

Marr, D. (1969): "A theory of cerebellar cortex". J. Physiol. 202, 437-470.

-------- (1970): "A theory of cerebral cortex". Proc. Royal Soc. London B 176, 161-234.

-------- (1971): "Simple memory: A theory for archicortex". Philos. Trans. Royal Soc. London 262, 23-81.

May, R.M. (1976): "Simple mathematical models with very complicated dynamics". Nature 261, 459-467.

McAllister, R.E., D. Noble and R.W. Tsien (1975): "Reconstruction of the electrical activity of cardiac Purkinje fibres". J. Physiol. (London) 251, 1-59.

McFadden, D. ed. (1979): "Neural Mechanisms and Behavior". Springer-Verlag.

McGregor, R.J. and Oliver (1974): "Model of repetitive firing in neurons". Kybernetik 16, 53-64 and 79-89.

-------------- and E.R. Lewis (1977): "Neural Modelling". Plenum Press, New York.

Minorsky, N. (1962): "Nonlinear Oscillators". Van Nostrand.

Minsky, M.L. and S. Papert (1969): "Perceptrons: An introduction to computational geometry". MIT Press, Cambridge.

Moore, G.P., D.H. Perkel and J.P. Segundo (1966): "Statistical analysis and functional interpretation of neuronal spike data". Am. Rev. Physiol. 28, 493-522.

----------- , J.P. Segundo and D.H. Perkel (1963): "Stability patterns in interneuronal pacemaker regulation". Proc. San Diego Symp. Biomed. Eng., 184-193.

Nagumo, J. and S. Sato (1971): "On a response characteristic of a mathematical neuron model". Kybernetic 10, 155-164.

Nakano, K. (1972): "Associatron - A model of associative memory". IEEE Trans. Syst., Man and Cybern., SMC-2, 380-388.

Narendra, K.S. and M.A.L. Thathachar (1974): "Learning automata: a survey". IEEE Trans. Syst., Man and Cybern., SMC-4, 323-334.

Neu, (1979): "Coupled chemical oscillators". SIAM J. Appl. Math. 37, 307-315.

Nietzsche, F. (1901): "Wille zur Macht", edited by E. Förster-Nietzsche Kröner, Leipzig.

Nilsson, N.J. (1965): "Learning machines". McGraw-Hill.

Parnas, I., D. Amstrong and F. Strumwasser (1974): "Prolonged excitatory and inhibitory synaptic modulation of a bursting pacemaker neuron". J. Neurophysiol. 7, 594-608.

Pavlidis, T. (1973): "Biological Oscillators: Their mathematical analysis". Academic Press.

------------ (1982): "Algorithms for graphics and image processing". Springer-Verlag.

------------ and H.N. Pinsker (1976): "Oscillator theory and neurophysiology". Fed. Proc. 36, 2033-2059.

Pavlov, I.P. (1927): "Conditioned Reflexes". Oxford Univ. Press.

Paz, O. (1975): "Pasado en claro". Fondo de Cultura Económica.

Perkel, D.H., J.H. Schulman, T.H. Bullock, G.P. Moore and J.P. Segundo (1964): "Pacemaker neurons: Effects of regularly spaced synaptic input". Science 145, 61-63.

Peterson, E.L. and R.L. Calabrese (1982): "Dynamic analysis of a rhythmic neural circuit in the leech Hirudo medicinalis". J. Neurophysiol. 47, 2.

Pinsker, H.M. (1977a): "Aplysia bursting neurons as endogenous oscillators. I - Phase response curves for pulsed inhibitory synaptic input". J. Neurophysiol. 40, 527-543.

------------- (1977b): "Aplysia bursting neurons as endogenous oscillators. II - Synchronization and entrainment by pulsed inhibitory synaptic input". J. Neurophysiol. 40, 544-556.

------------- and J. Ayers (1983): "Neuronal Oscillators". Chapter 9 in "Neurobiology", edited by W.D. Willis, Churchill Livingstone Inc.

------------- and E.R. Kandel (1977): "Short-term modulation of endogenous bursting rhythms by monosynaptic inhibition in Aplysia neurons: effects of contingent stimulation". Brain Res. 125, 51-64.

Polya, G. (1945): "How to solve it". Princeton Univ. Press.

Popper, K.R. and J.C. Eccles (1977): "The Self and its Brain". Springer-Verlag.

Ramón y Cajal, S. (1897): "Reglas y consejos sobre investigación científica (Los tónicos de la voluntad)". Lecture of entrance in the Real Academia de Ciencias Exactas, Físicas y Naturales. In "Obras literarias completas", Aguilar 1969.

------------------ (1899): "Textura del sistema nervioso del hombre y los vertebrados". Imprenta y Librería de Nicolás Moya, Madrid.

------------------ (1921): "Charlas de café (Pensamientos, anécdotas y confidencias)". In "Obras literarias completas", Aguilar 1969.

Ramos, A., E.L. Schwartz and E.R. John (1976a): "An Examination of the participation of Neurons in Readout from Memory". Brain Res. Bull. 1, 77-86.

--------- , ------------ , ---------- (1976b): "Stable and Plastic Unit Discharge Patterns during Behavioral Generalization". Science 192, 393-396.

Rescigno, A., R.B. Stein, R.L. Purple and R.E. Poppele (1970): "A neuronal model for the discharge patterns produced by cyclic inputs". Bull. Math. Biophys. 32, 337-353.

Rescorla, R.A. and A.R. Wagner (1972): "A theory of Pavlovian conditioning: Variations in the effectiveness of reinforcement and non-reinforcement". In "Classical conditioning. II - Current research and theory", edited by A.H. Black and W.F. Prokasy, Appleton-Century-Crofts.

Rosenblatt, F. (1962): "Principles of neurodynamics". Spartan Books.

Rosenzweig, M.R. and E.L. Bennet eds. (1976): "Neural Mechanisms of Learning and Memory". MIT Press.

Sábato, E. (1951): "Hombres y engranajes". Emecé, Buenos Aires.

---------- (1953): "Heterodoxia". Emecé, Buenos Aires.

Sato, S. (1972): "Mathematical Properties of Responses of a Neuron Model. A System as a Rational Number Generator". Kybernetik 11, 208-216.

Schwindt, P.C. and W.H. Calvin (1973): "Equivalence of synaptic and injected current in determining the membrane potential trajectory during motor neuron rhythmic firing". Brain Res. 59, 389-394.

Searle, J. (1980): "Minds, Brains, and Programs". In "The Behavioral and Brain Sciences", vol. 3, Cambridge Univ. Press.

Segundo, J.P. (1979): "Pacemaker synaptic interactions: Modelled locking and paradoxical features". Biol. Cybern. 35, 55-62.

-------------- and A.F. Kohn (1981): "A model of excitatory synaptic interactions between pacemakers. Its reality, its generality and the principles involved". Biol. Cybern. 40, 113-126.

-------------- and D.H. Perkel (1969): "The nerve cell as an analyzer of spike trains". In "The interneuron", UCLA Forum in Medical Sciences, num. 11, edited by M.A.B. Brazier, Univ. California Press, Los Angeles.

Selverston, A. (1976): "A Model System for the Study of Rhythmic Behavior". In "Simpler Networks and Behavior", edited by J.C. Fentress, Sinauer Associates.

Shepherd, G.M. (1974): "The Synaptic Organization of the Brain". Oxford Univ. Press.

Singer, W. (1977): "Control of Thalamic Transmission by Corticofugal and Ascending Reticular Pathways in the Visual System". Physiol. Rev. 57, num. 3, 386-420.

Skinner, B.F. (1938): "The Behavior of Organisms: An Experimental Analysis". Appleton Century.

Smith, D.R. and G.K. Smith (1965): "A statistical analysis of the continual activity of single cortical neurons in the cat anaesthetized isolated forebrain". Biophys. J. 5, 47-74.

Smith, G.K. and D.R. Smith (1964): "Spike activity in cerebral cortex". Nature 202, 253-255.

Sokolove, P.G. (1972): "Computer simulation of after-inhibition in crayfish slowly-adapting stretch receptor neuron". Biophys. J. 12, 1429-1451.

Spinelli, D.N. (1970): "OCCAM: A computer model for a content addressable memory in the central nervous system". In "The biology of memory", edited by K. Pribram and D. Broadbent, Academic Press.

Spira, P.M. (1969): "The time required for group multiplication". J. ACM 16, 235-243.

----------- and M.A. Arbib (1967): "Computation times for finite groups, semigroups, and automata". Proc. IEEE 8th Annual Symp. Switching and Automata Theory, 291-295.

Štefan, P. (1977): "A theorem of Sarkovskii on the existence of periodic orbits of continuous endomorphisms of the real line". Commun. Math. Phys. 54, 237-248.

Stein, P.G. (1974): "Neural control of interappendage phase during locomotion". American Zoologist 14, 1003-1016.

Stein, R.B., A.S. French and A.V. Holden (1972): "The frequency response, coherence, and information capacity of two neuronal models". Biophys. J. 12, 295-322.

Stevens, Ch.F. (1979): "La Neurona". Investigación y Ciencia", num. 38.

Strumwasser, F. (1965): "The demonstration and manipulation of a circadian rhythm in a single neuron". In "Circadian clocks", edited by J. Aschoff, North-Holland, 442-462.

----------------- (1967): "Types of information stored in single neurons". In "Invertebrate Nervous Systems. Their significance for mammalian neurophysiology", edited by C.A.G. Wiersma. Univ. Chicago Press.

----------------- (1974): "Properties of neurons and their relationship to concepts of plasticity". In "Neural mechanisms of learning and memory", edited by M.R. Rosenzweig and E.L. Bennet, MIT Press.

Sutton, R.S. and A.G. Barto (1981): "Toward a Modern Theory of Adaptive Networks: Expectation and Prediction". Psychol. Rev. 88, num. 2, 135-170.

Szentágothai, J. and M.A. Arbib (1974): "Conceptual models of neural organization". Neurosc. Res. Prog. Bull. 12, num. 3.

Takeuchi, A. and S. Amari (1979): "Formation of Topographic Maps and Columnar Microstructures in Nerve Fields". Biol. Cybern. 35, 63-72.

Teorell, T. (1971): "A biophysical analysis of mechano-electrical transduction". In "Handbook of Sensory Physiology. I - Principles of Receptor Physiology", edited by W.R. Loewenstein, Springer-Verlag.

Teyler, T. ed. (1978): "Brain and Learning". Greylock Publishers.

Thatcher, R.W. and E.R. John (1977): "Functional Neuroscience. Vol. 1 - Foundations of Cognitive Processes". Lawrence Erlbaum Associates.

-------------- and D.P. Purpura (1972): "Maturational status of inhibitory and excitatory synaptic activities of thalamic neurons in neonatal kitten". Brain Res. 44, 661-665.

-------------- and ------------ (1973): "Postnatal development of thalamic synaptic events underlying evoked recruiting responses and electrocortical activation". Brain Res. 60, 21-34.

Thompson, R.F., M.M. Patterson and T.J. Teyler (1972): "The neurophysiology of learning". Ann. Rev. Psychol. 23, 73-104.

Torras, C. (1981): "Neural model for the recognition of temporal patterns of stimulation". Master Thesis, Computer and Information Sciences Dept., University of Massachusetts, Amherst.

---------- (1982a): "A pacemaker model which displays phase-shifts, entrainment and plasticity in its firing pattern". Symp. "Ramón y Cajal" on Neuroscience.

---------- (1982b): "Modelling and simulation of a plastic pacemaker neuron". 2nd World Conf. on Math. at the Service of Man, 633-640.

---------- (1983): "Correspondència 2D-1D: Aplicació al càlcul de la TFD i la convolució circular". Tesi de Llicenciatura, Facultat de Matemàtiques, Universitat de Barcelona.

---------- (1985): "Pacemaker Neuron Model with Plastic Firing Rate: Entrainment and Learning Ranges". Biol. Cybern. 52, to appear.

Tosney, T. and G. Hoyle (1977): "Computer-controlled learning in a simple system". Proc. Royal Soc. London B 195, 365-393.

Tsetlin, M.L. (1973): "Automaton theory and modeling of biological systems". Academic Press.

Tsukahara, N. and M. Kawato (1982): "Dynamic and plastic properties of the brain stem neuronal networks as the possible neuronal basis of learning and memory". In "Competition and Cooperation in Neural Nets", edited by S. Amari and M.A. Arbib, Lecture Notes in Biomath. 45, Springer-Verlag.

Tuckwell, H.C. (1978): "Neuronal interspike time histograms for a random input model". Biophys. J. 21, 289-290.

Uttley, A.M. (1970): "The informon: A network for adaptive pattern recognition". J. Theor. Biol. 27, 31-69.

------------- (1975): "The informon in classical conditioning". J. Theor. Biol. 49, 355-376.

------------- (1979): "Information transmission in the nervous system". Academic Press, London.

van der Pol, B. (1926): "On relaxation oscillators". Phil. Mag. 2, 978-992.

Vinogradov, I. (1977): "Fundamentos de la teoría de números". Ed. MIR, Moscú.

Vinogradova, O. (1970): "Registration of information and the limbic system". In "Short-term Changes in Neuronal Activity and Behavior", edited by G. Horn and R.A. Hinde, Cambridge Univ. Press.

Vibert, J.F. and J.P. Segundo (1979): "Slowly adapting stretch-receptor organs: Periodic stimulation with and without perturbations". Biol. Cybern. 33, 81-95.

von Baumgarten, R. (1970): "Plasticity in the nervous system at the unitary level". In "The neurosciences: second study program", edited by F.O. Schmitt, Rockefeller Univ. Press, 260-271.

von der Malsburg, C. (1973): "Self-organization of orientation sensitive cells in the striate cortex". Kybernetik 14, 80-100.

Walloe, L., J.K.S. Jansen and K. Nygard (1969): "A computer simulated program of a second order sensory neuron". Kybernetik 6, 130-140.

Widrow, B., N.K. Gupta and S. Maitra (1973): "Punish/reward: learning with a critic in adaptive threshold systems". IEEE Trans. Syst., Man and Cybern., SMC-5, 455-465.

----------- and M.E. Hoff (1960): "Adaptative switching circuits". IRE WESCON Convention Rec. 4, 96-104.

Wigström, H. (1973): "A neuron model with learning capability and its relation to the mechanisms of association". Kybernetik 12, 204-215.

Willshaw, D.J. and C. von der Malsburg (1976): "How patterned neural connections can be set up by self-organization". Proc. Royal Soc. London B 194, 431-445.

Wine, J. and F. Krasne (1978): "The cellular analysis of invertebrate learning". In "Brain and Learning", edited by T. Teyler, Greylock Publishers.

Winfree, A.T. (1980): "The Geometry of Biological Time". Springer-Verlag.

Winograd, S. (1965): "On the time required to perform addition". J. ACM 14, 793-802.

Woody, C.D. (1982): "Memory, learning and higher function: A cellular view". Springer-Verlag.

Woolacott, M. and G. Hoyle (1977): "Neural events underlying learning in insects: Changes in pacemaker". Proc. Royal Soc. London B 195, 599-620.

Wyman, R.J. (1977): "Neural generation of the breathing rhythm". Ann. Rev. Physiol. 39, 417-448.

Yoshizawa, S. (1982): "Periodic pulse sequences generated by an analog neuron model". In "Competition and Cooperation in Neural Nets", edited by S. Amari and M.A. Arbib, Lecture Notes in Biomath. 45, Springer-Verlag.

------------- , H. Osada and J. Nagumo (1982): "Pulse sequences generated by a degenerate analog neuron model". Biol. Cybern. 45, 23-33.

Zeigler, B.P. (1976): "Theory of Modelling and Simulation". Wiley, New York.